PHARMACEUTICAL CHEMISTRY

PHARMACEUTICAL CHEMISTRY

Therapeutic Aspects of Biomacromolecules

CHRISTINE M. BLADON

Queen Mary, University of London

JOHN WILEY & SONS, LTD

Other Wiley Editorial Offices

John Wiley & Sons, Inc., 605 Third Avenue,
New York, NY 10158-0012, USA

Wiley-VCH Verlag GmbH, Pappelallee 3,
D-69469 Weinheim, Germany

John Wiley & Sons (Australia) Ltd, 33 Park Road, Milton,
Queensland 4064, Australia

John Wiley & Sons (Asia) Pte Ltd, 2 Clementi Loop #02-01,
Jin Xing Distripark, Singapore 0512

John Wiley & Sons (Canada) Ltd, 22 Worcester Road,
Rexdale, Ontario M9W 1L1, Canada

British Library Cataloguing in Publication Data

A catalogue record for this book is available from the British Library

ISBN 0 471 49636 7 (cased)   0 471 49637 5 (paperback)

Typeset in $10\frac{1}{2}/12\frac{1}{2}$ Sabon by Techset Composition Ltd, Salisbury, Wilts
Printed and bound in Great Britain by TJ International Ltd., Padstow, Cornwall
This book is printed on acid-free paper responsibly manufactured from sustainable forestry
in which at least two trees are planted for each one used for paper production.

Contents

Preface

The aim of this book is to introduce some of the strategies involved in the development and production of a range of clinically useful biomacromolecules and to describe some current advances which may lead to the design of future therapeutics. The work is intended to be an outline of why peptides, proteins, nucleic acids and oligosaccharides are developed as pharmaceuticals and how they are produced using techniques such as synthetic organic chemistry, molecular biology and biotechnology. Examples have been selected in order to emphasize the molecular basis of specific disease processes.

This book is based on a series of lectures I gave to final-year students on the Pharmaceutical Chemistry degree course at the University of Kent. As with many areas of science, the biomedical field uses specialized language and it can be difficult for students with a chemistry or biochemistry background to learn about new developments in an unfamiliar field. I have therefore attempted to predigest material to make it more understandable and hope that, with the inclusion of a glossary and appendix of basic biochemistry, the reader will be able to interpret and assess future developments in the treatment, and perhaps prevention, of medical conditions. A selection of books and recent reviews is provided at the end of each chapter to direct readers to more comprehensive coverage of the topics referred to in the main text. Key original papers are also cited so as to maintain an historical perspective.

I am indebted to my husband, Dr Peter Wyatt, my father, Dr Peter Bladon, colleagues Drs Alethea Tabor and Marina Resmini, and various referees who have read all or part of the manuscript and made many invaluable suggestions for its improvement. I hope that any mistakes that may have crept in are both minor and minimal and I accept full responsibility for them. Preparation of the manuscript has taken considerable time, not least because at times it seemed that the early months of motherhood and writing were incompatible activities. I am particularly grateful to my husband for his understanding, encouragement and practical assistance throughout the writing process and it is to him that I extend my deepest gratitude.

Christine Bladon

Abbreviations

ADA	Adenosine deaminase
AIDS	Acquired immune deficiency syndrome
APC	Antigen-presenting cell
ASO	Antisense oligonucleotide
Boc	t-Butoxycarbonyl
Bu^t	t-Butyl
Bzl	Benzyl
C	Constant region (of antibody)
CD	Cluster of differentiation
cDNA	Complementary DNA
CDR	Complementarity determining region
(k)Da	(kilo) Dalton
DNA	Deoxyribonucleic acid
dNTP	Deoxynucleotide triphosphate
E. coli	*Escherichia coli*
ELISA	Enzyme-linked immunosorbent assay
Fab	Fragment antibody binding
Fc	Fragment crystalline (of antibody)
Fmoc	9-Fluorenylmethoxycarbonyl
Fv	Fragment variable (of antibody)
HAMA	Human anti-mouse antibody
HIV	Human immunodeficiency virus
IL	Interleukin, e.g. IL-2, Interleukin 2
IL-2R	Interleukin 2 receptor
mAb	Monoclonal antibody
MHC	Major histocompatibility complex
mRNA	Messenger RNA
NK	Natural killer cell

PCR	Polymerase chain reaction
PEG	Poly(ethylene glycol)
rDNA	Recombinant DNA
RNA	Ribonucleic acid
RNase H	Ribonuclease H
scFv	Single chain fragment variable (of antibody)
SCID	Severe combined immunodeficiency
TAA	Tumour-associated antigen
T_c	Cytotoxic T cell
TCR	T-cell receptor
TFA	Trifluoroacetic acid
T_h	Helper T cell
tPA	Tissue plasminogen activator
V	Variable region (of antibody)
V_H	Variable region of antibody heavy chain
V_L	Variable region of antibody light chain
Z	Benzyloxycarbonyl

1

Introduction

1.1 Overview

An understanding of the molecular basis of a disease is a key element in identifying an appropriate treatment. Abnormalities in the production of polypeptides, nucleic acids and carbohydrates frequently lead to diseases which, in turn, can be treated by administration of these classes of chemical compound. This type of compound is what this book is about, although the 'macro' in its title is interpreted rather loosely.

Some of the most well-known, although rare, diseases such as haemophilia and severe combined immunodeficiency (SCID) disorders are caused by mutations in genes encoding particular enzymes. While such genetic disorders can be treated by replacing the deficient enzymes, identification of, and the ability to manipulate the genes responsible, is opening up new avenues for treatment. The immune system is affected in a great variety of diseases, ranging from the genetically determined SCID to environmentally determined conditions such as bacterial and viral infections. An understanding of the basic mechanisms of the immune system has helped in the immunoprophylaxis of, for example, the virus that causes poliomyelitis. The autoimmune diseases of rheumatoid arthritis and multiple sclerosis lie between the two extremes of the immune spectrum. Both genetic and environmental factors are thought to be involved in the pathogenesis of autoimmunity – a genetic defect may render an individual more susceptible to an initial environmental trigger. The more complex nature of the multifactorial diseases makes them very much more difficult to treat than the unifactorial disorders. Many forms of cancers are difficult to treat as again the aetiology is thought to involve the interaction of genetic and environmental factors. However, with increasing knowledge of the cellular and molecular biology of cancer, new therapeutic strategies are being developed. A brief analysis of the treatment for gigantism, dwarfism and the SCID disorder, adenosine deaminase (ADA) deficiency, will serve to illustrate some of the above points which are discussed more fully in subsequent chapters.

1.1.1 Endocrine disorders

Gigantism and some types of dwarfism are associated with dysfunction of the hypothalamus-pituitary endocrine system. Insufficient secretion by the pituitary of the protein growth hormone during childhood leads to small stature, although this condition can be treated by the administration of synthetic growth hormone; the best results are obtained when the treatment is started as early as possible. On the other hand, overproduction of the hormone, often the result of a tumour, results in increased growth and gigantism. Furthermore, if the tumour compresses neighbouring tissue the secretion of other pituitary hormones can be affected, thus delaying puberty with the result that the growth plates at the end of the long bones remain immature and continue to grow, leading to the attainment of an even greater ultimate height. In the past, gigantism generally resulted in a shortened life span. The Irish giant Charles Bryne, for example, was born in 1761, grew to a height of 7 ft 10 in (2.39 m) and died in his early 20s (Figure 1.1). Fortunately, gigantism can also now be diagnosed before it becomes so advanced and growth hormone concentrations can be reduced to acceptable levels by treating the tumours using surgery, radiotherapy and/or drug treatment. When these same tumours arise after puberty, at a time when the growth plates have matured and the bones have stopped growing, the condition of acromegaly develops. Symptoms of this syndrome develop over a number of years; skin and bones thicken but the latter do not lengthen, facial features become distorted, hands and feet become enlarged, and many internal organs increase in size (Figure 1.2). If left untreated, the acromegalic individual usually dies of cardiovascular disease. The condition is commonly treated with the drug octreotide (**1.1**), a synthetic analogue of the growth hormone inhibitor somatostatin (**1.2**), either alone or in combination with surgery.

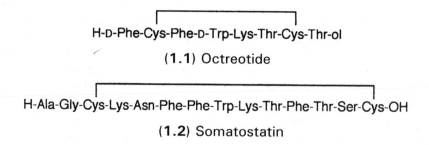

H-D-Phe-Cys-Phe-D-Trp-Lys-Thr-Cys-Thr-ol

(**1.1**) Octreotide

H-Ala-Gly-Cys-Lys-Asn-Phe-Phe-Trp-Lys-Thr-Phe-Thr-Ser-Cys-OH

(**1.2**) Somatostatin

1.1.2 Severe combined immunodeficiency disorders

These are a group of diseases in which both branches of the immune response (i.e. cellular immunity and antibody production) are defective. Children born with a non-existent immune system have no defence against pathogens and

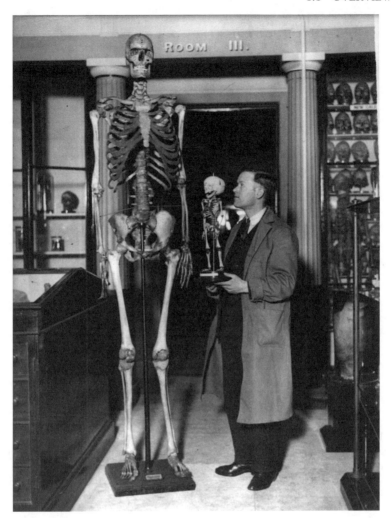

Figure 1.1 The skeleton of Charles Bryne is now in the Hunterian Museum, The Royal College of Surgeons, London, and is shown alongside a man of average stature – also shown is the skeleton of Caroline Crachami, the Sicilian dwarf. She was born weighing one pound (450 g) and measured 7 in (18 cm) in length, and when she died in 1824 she was approximately 22.5 in (52 cm) tall and weighed 5–6 lb (2.27–2.73 kg) – scarcely different from the size and weight of a normal child at birth. Reputed to be nearly nine years old at the time of her death, recent examination of her teeth suggests that she was about three years old. Reproduced by permission of the Royal College of Surgeons, London.

even minor infections can be life-threatening. Children with the most severe forms of severe combined immunodeficiency (SCID) are forced to live in protective isolation ('sterile bubbles') and cannot enjoy normal growth and development. In the SCID disease caused by ADA deficiency, the immune system fails to develop normally because of the accumulation of ADA substrates which are toxic to lymphocytes, the cells of the immune system.

Figure 1.2   Photographic record of the progression of acromegaly: facial features of an acromegalic individual at 9, 16, 33 and 52 years of age. The woman at age 56 died from congestive heart failure. Reprinted from "Acromegaly, Diabetes, Hypermetabolism, Proteinuria and Heart Failure," *American Jornal of Medicine*, **20**, 133–144, 1956, with permission of Excerptu Medica Inc.

The disease can be kept at bay by continuous administration of the enzyme ADA. A better therapeutic approach is to rebuild the immune system from within. Bone marrow transplantation is a viable option if a fully-matched donor can be found but the recipient has to take immunosuppressive drugs to prevent rejection. Transfer of a normal gene into the immune cells would be a more complete cure and ADA-deficient individuals have been among the first to benefit from this, albeit still experimental, gene therapy option.

1.2   Biomacromolecules in the Treatment of Human Disease

Therapeutic peptides and proteins are used to treat the metabolic or biochemical consequences of the basic defects underlying human disease, rather than being aimed directly at the causative agents themselves. The variety of peptide and protein therapeutics range from hormones and enzymes to vaccines and monoclonal antibodies, and many diseases once thought incurable can now be successfully treated. Targeting defects in nucleic acids or affecting their translation is the province of oligonucleotide-based therapeutics. These latter compounds are still largely at the developmental stage and are not in routine clinical use, although some have entered clinical trials. Oligosaccharides play a critical role in a wide range of cellular recognition processes and as such they have considerable potential as therapeutic agents. However, the structural complexity of these compounds and difficulties in their syntheses have caused the oligosaccharides to be an under-represented class of therapeutic bio-macromolecules.

Pharmaceutical biomacromolecules can be used to treat human disease in three main ways, as follows:

1. *Direct replacement* of a protein, the deficiency of which causes disease. Well-known examples of such diseases include diabetes mellitus which is due to a lack of insulin, the blood-clotting disorder haemophilia as a result of a deficiency in factor VIII, and dwarfism, which is caused by insufficient growth hormone production during childhood. These conditions can each be effectively treated by the administration of the relevant protein frequently for the lifetime of the individual or, in the case of pituitary dwarfism, during the critical years of development.

   Haemophilia and some forms of dwarfism are single-gene disorders and the analogous oligonucleotide treatment would be to administer a corrected copy of the defective gene. Gene-transfer therapy is not currently a routine clinical treatment although the feasibility of the concept has been proved.

2. *Symptomatically*, to achieve highly selective, short-term intervention for modifying pathogenic processes. For example, blood clots which can precipitate heart attacks and strokes may be degraded with the enzyme tissue plasminogen activator and thus prevent recurrence of the condition. A *potential* treatment for diseases such as cancer which result from incorrect gene function or expression of specific genes is antisense DNA therapy. In this approach, oligonucleotides selectively block the disease-causing genes, thereby inhibiting the production of disease-associated proteins. The feasibility of this type of treatment is currently being evaluated and the results auger well for the clinical licensing of antisense oligonucleotide therapeutics.

3. *Prophylactically*, to confer immunological protection against a variety of pathogens. Many diseases such as poliomyelitis, measles and whooping cough are now preventable by immunizing with vaccines composed of inactivated or attenuated strains of the organism or virus. With some

pathogenic organisms it is now possible to identify the discrete molecular structures that are responsible for stimulation of the immune system. Once identified, these epitopes, most frequently surface proteins or, less commonly, oligosaccharide chains, can be isolated or synthesized and developed into a vaccine.

Peptides and proteins are the biomacromolecules most frequently used in the treatment of human disease and their widespread and varied application is reflected in the proportion of this book which is devoted to these compounds. The exciting clinical potential of gene therapy and antisense agents is discussed in the chapter on oligonucleotides. The text concludes with a look at some clinically useful oligosaccharides, a less mature but nevertheless valuable class of therapeutic biomacromolecules.

2

Endogenous Peptides and Proteins

2.1 Overview

Endogenous peptides and proteins are those which are produced within the human body. Using such molecules in therapy often has the obvious advantage that, by their very nature, they will effect the desired biological response. Examples include hormones, components in enzyme cascades, cytokines, growth factors and antibodies. This diversity of product types reflects the wide range of conditions that can be treated with endogenous peptides and proteins. Usually, there are no adverse effects associated with the administration of endogenous compounds which normally circulate in the blood, such as insulin and other hormones. Difficulties that arise do so because under physiological conditions peptides and proteins have short half-lives and are generally not orally active.

Endogenous proteins have traditionally been obtained from human or animal tissue or fluids, although more recent sources include semi-synthesis and recombinant DNA (rDNA) production. They can be used in the form found in nature or their structures can be modified to produce a more active substance or a substance which is more specifically targeted or one which can be administered more easily (see Chapter 3). For example, peptides which are metabolically unstable and are poorly absorbed orally do not make good drug candidates and, to be clinically useful, a mimetic based on the natural substance is developed in order to exploit the desirable properties. Proteins can be redesigned at the genetic level by site-directed mutagenesis or derivatized chemically to improve their therapeutic potential. Some examples of peptides and proteins currently in use as therapeutic agents are given in Table 2.1.

Table 2.1 A selection of peptides and proteins currently used as therpeutic agents

Peptide/Protein	Natural sources[a]	Main biological effects	Therapeutic applications	Sources of pharmaceutical product
Insulin (51 *Amino acids*)	Pancreas	Regulation of carbohydrate metabolism	Diabetes mellitus	Semi-synthesis, rDNA
Glucagon (29 *Amino acids*)	Pancreas	Mobilization of hepatic glucose	Reverse-insulin-induced hypoglycaemia	Animal tissue
Growth hormone (191 *Amino acids*)	Anterior pituitary	Promotes growth of tissues (deficiency causes dwarfism, overproduction causes gigantism, (acromegaly in adults))	Dwarfism	rDNA
Somatostatin (14 *Amino acids*)	Hypothalamus, pancreas	Inhibits release of insulin and glucagon from pancreas, inhibits release of growth hormone from anterior pituitary	Acromegaly, diabetes mellitus	Chemical synthesis; analogues with longer duration of action are preferred as therapeutics
Factor VIII (2332 *Amino acids*)	Plasma	Clotting factor	Haemophilia	Blood fraction, rDNA
Tissue plasminogen activator (527 *Amino acids* (ca. 70 kDa))	Plasma	Catalyses conversion of plasminogen to plasmin	Angina, stroke, heart attack, wound healing	rDNA, milk of transgenic goats

Urokinase (366 *Amino acids* (ca. 54 kDa))	Plasma	Catalyses conversion of plasminogen to plasmin	Angina, stroke, heart attack, wound healing	Human urine, tissue culture of human kidney cells
Streptokinase (~ 45 kDa)	Microbial protein	Activates plasminogen	Angina, stroke, heart attack, wound healing	Culture filtrate of *Streptococcus haemolyticus*
α_1-Antitrypsin (394 *Amino acids* (ca. 52 kDa))	Serum	Inhibits protease elastase	Emphysema	Milk of transgenic sheep
Oxytocin (9 *Amino acids*)	Posterior pituitary	Contraction of uterus and milk ducts	Induction and maintenance of labour	Chemical synthesis; a more active analogue sometimes used as an alternative to the native hormone as it exhibits a higher degree of potency and longer circulatory half-life
Vasopressin (9 *Amino acids*)	Posterior pituitary	Regulates reabsorbtion of water by kidneys	Diabetes insipidus (rare form of diabetes due to deficiency of vasopressin)	Chemical synthesis; more active analogues often used clinically as they exhibit greater anti-diuretic activity and more prolonged period of action

(*continued*)

Table 2.1 A selection of peptides and proteins currently used as therapeutic agents (*continued*)

Peptide/Protein	Natural sources[a]	Main biological effects	Therapeutic applications	Sources of pharmaceutical product
Luteinizing hormone-releasing hormone (LHRH) (*10 Amino acids*)	Hypothalamus	Stimulates synthesis of follicle-stimulating hormone (FSH), luteinizing hormone (LH), in anterior pituitary; hormonal regulation of the sex steroid hormones	Infertility and contraception in women, prostate cancer in men	Chemical synthesis; analogues are usually more potent and have longer duration of action and are used clinically, e.g. buserelin and goserelin in the treatment of prostate cancer
Gonadotrophins (FSH, LH, chorionic gonadotrophin (CG)) (*FSH*, 34 kDa; *LH*, 28.5 kDa; *CG*, 36.6 kDa)	FSH and LH (anterior pituitary), CG (placenta)	Regulation of sex steroid hormones	Infertility – used to induce superovulation	Urine of postmenopausal women (LH, FSH), urine/placenta of pregnant women (CG)
Immunoglobulins and antisera (ca. 150 kDa)	Plasma	Neutralization of toxic substances	Passive immunization	Donated blood (polyclonal antibodies), cell culture (monoclonal antibodies)
Vaccines	N/a	Prime immune system	Active immunization	Inactivated or attenuated strains of virus/bacteria, rDNA of surface antigens
Cytokines	T cells	Immunomodulators	Immune-mediated disorders, viral infections, certain tumours	rDNA, cell culture

[a]N/a, not applicable

2.2   Isolation from Natural Tissues and Fluids

Until recently, endogenous proteins used for replacement therapy were only available from a few biological sources, for example, insulin was obtained from pancreatic tissue of slaughterhouse animals and blood factors were isolated from donated plasma. The level of protein found in body tissues and extracellular fluids is normally very low and large quantities of the source material are required to obtain appreciable amounts of the desired protein. For example, the amount of insulin obtained from the pancreatic tissue of three pigs only satisfies the requirements of one diabetic patient for about 10 days. Nevertheless, a number of therapeutic proteins are still produced by extractive methods from biological fluids despite the advent of rDNA technology. Therapeutic peptides are normally less than 30 residues in length and, as a rule, it is more economical to produce these compounds by chemical synthesis rather than by extractive methods from natural sources. To put this in context, in the original isolation, purification and characterization studies of the hypothalamic hormone thyrotrophin-releasing hormone (pGlu-His-Pro-NH$_2$) in the 1960s, almost 4 t of pig hypothalamic tissue yielded only approximately 1 mg of pure peptide.

The purity and homogeneity of the protein once it has been isolated from the source material is established by using a combination of chromatographic and analytical techniques. The presence of impurities can render the protein immunogenic, i.e. appear foreign to the human body. Immune reactions and the production of antibodies which neutralize the biological effect of the protein can also be caused if the protein administered to a patient is not identical to the human form. Animal proteins frequently differ from the human counterpart in amino acid sequence. For example, in human insulin the C-terminal residue of the B-chain is threonine, whereas the corresponding residue in the porcine protein is alanine. Many of the larger proteins have oligosaccharide chains attached to certain asparagine, threonine and/or serine residues, and the animal protein may feature a different glycosylation pattern to that of the human material. The presence of neutralizing antibodies can be a problem as increased doses are needed to overwhelm the antibodies and this makes the treatment not only more expensive but it can also complicate the management of the disease.

A further problem with the therapeutic use of proteins isolated from natural sources is the potential presence of pathogens in the raw material. Until 1985, children with pituitary dwarfism were treated with growth hormone obtained from human pituitaries at post-mortem. Tragically, many of the children treated subsequently developed Creutzfeldt–Jacob disease, a neurodegenerative disorder, because of infectious prion particles. Such particles are unique in that they contain no nucleic acid but an abnormal isoform of the prion protein. Transmission of the prions can now be avoided by the use of synthetic human growth hormone produced by rDNA techniques. More recently, blood transfusions contaminated with the human immunodeficiency virus (HIV) led to a large number of haemophiliac patients contracting acquired immune deficiency syndrome (AIDS). Great efforts are therefore made in the isolation and purification protocols in order to ensure that protein material destined for therapeutic use is safe.

2.3   Synthesis and Semi-Synthesis

Many polypeptide hormones are less than 30 residues in length; solution and solid-phase synthesis techniques can produce these substances economically on a large scale. The distinction between peptides and proteins is somewhat arbitrary but sequences consisting of less than 50 residues are referred to as *peptides*, whereas if the number of amino acids exceeds 50 then the compound is usually referred to as a *protein*. The chemical synthesis of proteins is not normally economic and substantial quantities of material are often obtained by rDNA technology (see Section 2.4). However, it can be cost-effective in some instances to combine synthetic techniques with extractive methodology and modify naturally occurring proteins. One example where this approach has been particularly successful is the preparation of human insulin from the insulin isolated from pigs. As noted above in Section 2.2, human insulin differs from porcine insulin by one amino acid, but this single residue in the latter protein can be replaced by using enzymatic techniques and thus effectively convert the porcine sequence into the human sequence. Further details of this procedure are described below in Section 2.6.1.

2.3.1   Peptide synthesis

The chemical synthesis of peptides originated in the early 20th century. At first, such syntheses were difficult and limited to simple amino acids. The discovery of the easily removable benzyloxycarbonyl group in 1932 and improvements in methods of amide-bond formation in the 1950s paved the way for the preparation of longer peptides and those incorporating polyfunctional amino acids. The first peptide hormone to be synthesized was oxytocin (2.1) carried out by du Vigneaud and collaborators in the 1950s. Oxytocin is a cyclic nonapeptide amide secreted by the hypothalamic-pituitary endocrine system which controls uterus contractions and breast milk ejection. The synthetic hormone is used in obstetric medicine to induce childbirth and to

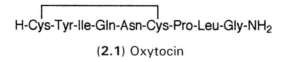

H-Cys-Tyr-Ile-Gln-Asn-Cys-Pro-Leu-Gly-NH$_2$

(**2.1**) Oxytocin

promote lactation in cases of faulty milk ejection. In the original synthesis, a segment condensation approach was adopted in which smaller peptide fragments were prepared and then linked together to give the target sequence. A subsequent synthesis by du Vigneaud still used the so-called 'classical' solution chemistry but the peptide chain was assembled by the addition of single residues starting from the C-terminus and building towards the N-terminus (Scheme 2.1). This stepwise-addition strategy uses N^α-protected-C^α-activated amino acids and is similar to present day synthetic protocols.

While recent advances in the peptide field may have improved individual components of synthetic procedures, du Vigneaud's stepwise oxytocin synthesis illustrates many of the key aspects involved in the preparation of peptides

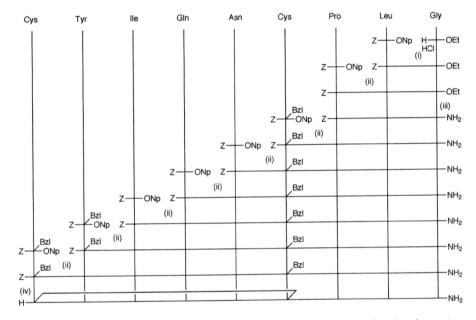

Scheme 2.1 Synthesis of oxytocin. The synthesis is presented by using shorthand notation. Vertical lines indicate amino acids, horizontal lines represent bonds and side-chain protecting groups are denoted by a short line at an angle of 45° to the right of the vertical line. Thus, in the first step leucine, N^α-protected with a Z group and activated as the *p*-nitrophenolate, was coupled to glycine with its carboxyl protected as an ethyl ester. The amino-terminal Z protecting group of the product from this first step, Z-Leu-Gly-OEt, was then removed and the resulting dipeptide H-Leu-Gly-OEt was coupled to proline which had Z protection and *p*-nitrophenyl activation. The first cysteine, asparagine, glutamine, isoleucine, tyrosine and the second cysteine were coupled in a similar manner to proline. The thiol group of the cysteine residues and the phenolic hydroxy group of tyrosine were protected as benzyl esters. In the final step, all of the protecting groups were removed and the disulphide bond was formed. *Note:* the C-terminal ethyl ester was converted to an amide following addition of the proline. Reaction conditions: (i) $Et_3N/CHCl_3$; (ii) HBr/AcOH on the protected amino component, then conversion to the free base form and reaction with the active ester; (iii) $NH_3/MeOH$; (iv) Na/NH_3 (l), then air oxidation and purification

today. When planning a synthesis three issues need to be considered, namely strategy, choice of protecting groups, and activation and coupling methods.

(1) Strategy

Stepwise chain assembly starting with the C-terminal residue requires the incorporation of each residue in a protected and activated form and removal of the α-amino protecting group after each chain-lengthening step (Scheme 2.2). Use of urethane protecting groups, exemplified by benzyloxycarbonyl in the oxytocin synthesis, prevents racemization during the peptide-bond forming reaction and renders this approach more practical than assembly in the N → C direction.

(2) Protection of functional groups

Formation of an amide bond unambiguously between the amino group of one amino acid and the carboxyl group of a second requires that all functional groups, except those actually involved in formation of the peptide bond, are protected. In the stepwise assembly strategy, protection of the α-amino function is short-term or temporary, and is removed after each cycle of amino acid addition, while all other functionality is masked throughout the synthesis and only removed after chain building is complete; this is known as long-term or permanent protection. A further requirement is that the short-term protecting groups must be removable under conditions which do not affect the long-term protecting groups and the chiral integrity of the growing peptide chain.

(1) α-Amino-protection

Urethane (alkoxycarbonyl) groups have dominated protection of the α-amino functionality. They have remained popular because they prevent racemization during the coupling step and also provide scope for selective cleavage. Fission of the parent O–R bond gives a carbamic acid which spontaneously decarboxylates to regenerate the parent amine (Scheme 2.3). By changing the nature of the R group, cleavage of the O–R bond can be achieved under a range of conditions. Benzyl (2.2, R = CH_2Ph) was the original urethane protecting group and is abbreviated by 'Z' after its inventor Zervas. Cleavage of the Z group can be effected with HBr in AcOH in an S_N2 mechanism or by catalytic hydrogenolysis. Two other particularly useful types of urethanes are the t-butoxycarbonyl (Boc, 2.3) and 9-fluorenylmethoxycarbonyl (Fmoc, 2.4) groups. The N-Boc group is stable to catalytic hydrogenolysis but is labile towards acids and is commonly removed with trifluoroacetic acid (TFA). The N-Fmoc group is

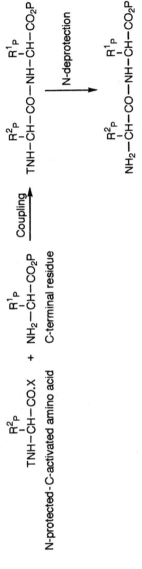

Scheme 2.2 Unambiguous formation of a peptide bond: T, short-term protecting group; P, long-term protecting group; R^1, R^2, side chains; X, activating group

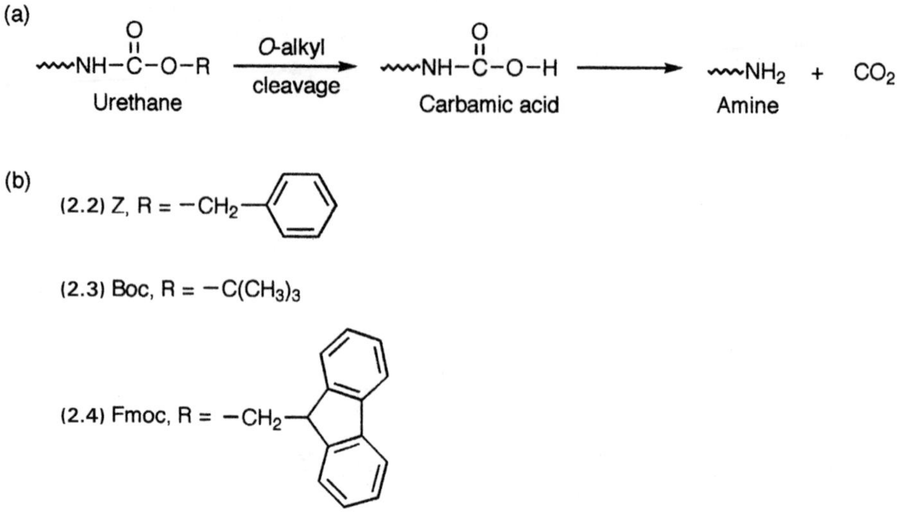

(a)

$$\text{wwNH}-\overset{\overset{\displaystyle O}{\|}}{C}-O-R \xrightarrow[\text{cleavage}]{\textit{O}\text{-alkyl}} \text{wwNH}-\overset{\overset{\displaystyle O}{\|}}{C}-O-H \longrightarrow \text{wwNH}_2 + CO_2$$

Urethane   Carbamic acid   Amine

(b)

(2.2) Z, R = $-CH_2-$⟨phenyl⟩

(2.3) Boc, R = $-C(CH_3)_3$

(2.4) Fmoc, R = $-CH_2-$⟨fluorenyl⟩

Scheme 2.3 Urethane protecting groups: (a) reaction scheme for deprotection via a carbamic acid; (b) structures of the most commonly used urethane protecting groups in peptide synthesis

stable to acidic reagents but is cleaved rapidly with secondary amines, with treatment by 20% piperidine in N,N-dimethylformamide (DMF) for 10 min being the standard deprotection conditions.

(2) α-Carboxyl-protection

Only the α-carboxyl group of the C-terminal residue requires permanent protection; the carboxyl functions of all subsequent amino acids are added in an activated form. The usual means of protection is esterification. Methyl and ethyl esters provide good carboxyl protection but they are rather too stable for peptide synthesis and harsh alkaline hydrolysis is required for their removal. The use of t-butyl and benzyl esters can avoid these rather vigorous deprotection conditions. The lability of benzyl and t-butyl esters generally parallels that of the Z and Boc groups respectively; benzyl esters are cleaved by catalytic hydrogenolysis or HBr in AcOH, while TFA removes t-butyl esters.

(3) Side-chain-protection

Besides the amino acids with additional amino and carboxyl groups (lysine, aspartic acid and glutamic acid), the hydroxy-containing amino acids (serine, threonine and tyrosine), together with cysteine, tryptophan, arginine and histidine, all contain reactive groupings which are likely to interfere with specific peptide bond formation and so require permanent protection (Table 2.2).

Table 2.2 Common side-chain protecting groups

Side-chain functionality of amino acid	Protecting group	Abbreviation[a]	Cleavage conditions
Arginine			
		NO_2	H_2, Pd–C
		Mts	HF
		Mtr	TFA[b]
		Pmc	TFA
Aspartic acid, Glutamic acid			
		OBzl	H_2, Pd–C
		OBu^t	TFA
Cysteine			
	S–CH$_2$-NH-CO-CH$_3$	Acm	HgCl$_2$ or I$_2$[c]
	S–C(CH$_3$)$_3$	Bu^t	TFA
	S–C(Ph)$_3$	Trt	TFA
Histidine[d]			
		Bom	HF or H_2, Pd–C
		Boc	TFA
		Trt	TFA
Lysine			
		Z	HF or H_2, Pd–C
		Boc	TFA

(continued)

Table 2.2   Common side-chain protecting groups (*continued*)

Side-chain functionality of amino acid	Protecting group	Abbreviations[a]	Cleavage conditions
Serine, threonine, tyrosine			
∿OH	∿O−CH$_2$Ph	Bzl	HF or H$_2$ Pd−C
	∿O−C(CH$_3$)$_3$	Bu^t	TFA
Tryptophan		For	HF^e
		Boc	TFA

[a]Mts, 2,4,6-Trimethylbenzenesulphonyl; Mtr, 4-methoxy-2,3,6-trimethylbenzenesulphonyl; Pmc, 2,2,5,7,8-pentamethylchroman-6-sulphonyl; Acm, acetamidomethyl; Trt, trityl; Bom, benzyloxy-methyl.

[b]Deprotection by TFA can be slow, especially in multiple Arg(Mtr)-containing peptides.

[c]I$_2$ cleavage results in formation of cystine (cysteine disulphide).

[d]Steric resistance in the reagent favours the τ-position, so that, e.g. trityl, is regioselective for the τ-nitrogen. The π-Bom derivative is prepared by substituting the τ-position first, then alkylating the π-nitrogen, with final cleavage of the τ-substituent.

[e]HF containing thiophenol, or alternatively the formyl group can be removed with piperidine.

In the oxytocin synthesis, the marked resistance of the benzyl group on the thiol function of cysteine to acids allowed for the temporary protection of the α-amino groups with Z because the latter can be cleaved with HBr in AcOH without liberating the SH group. Benzyl protection of the tyrosine phenolic group was, however, labile to acidolysis and was cleaved during removal of the tyrosine Z group. This is not an ideal situation and it is preferable to use combinations of temporary and permanent protecting groups which are complementary or 'orthogonal' to each other. Combinations which have proved particularly valuable in recent years are (a) N^α-Boc (temporary) plus benzyl-based side-chain protection (permanent) which are removable with TFA and catalytic hydrogenolysis or HF respectively, and (b) N^α-Fmoc combined with *t*-butyl-based side-chain protection which can be cleaved with piperidine and TFA, respectively (Figure 2.1).

(3)   Activation and coupling

Formation of an amide bond between two amino acids will only occur if the electrophilicity of the carbon atom of the carboxyl component is enhanced by

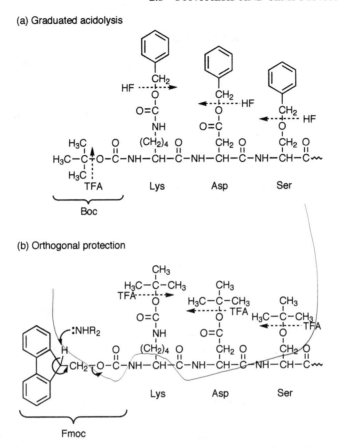

Figure 2.1 Commonly used protection schemes. (a) Graduated acidolysis – Temporary N^α-amino protection is provided by the Boc group, removed by TFA, while the permanent Bzl-based side-chain protecting groups are cleaved by HF. (b) Orthogonal protection – Temporary N^α-amino protection is provided by the Fmoc group, removed by the indicated base-catalysed β-elimination mechanism. Permanent Bu^t-based side-chain protecting groups are cleaved by TFA

the presence of an electron-withdrawing group. Such activation of the carboxyl group can be accomplished in a number of ways (Figure 2.2). In the oxytocin synthesis, *p*-nitrophenyl (ONp) esters were used as the acylating agents and this active ester method has proved a steadfast approach over the years. Aryl esters incorporating an electron-withdrawing group in the aromatic ring are generally stable crystalline derivatives of N^α-urethane protected amino acids and they are easily prepared. *p*-Nitrophenyl esters can be slow to react, sometimes requiring about 12 h for complete coupling, and they have largely been superseded by pentafluorophenyl (OPfp) esters which generally react within 1 h. Symmetrical anhydrides are highly reactive and an efficient method of acylation. However, there are disadvantages with this activation approach. First, the anhydrides must be prepared immediately

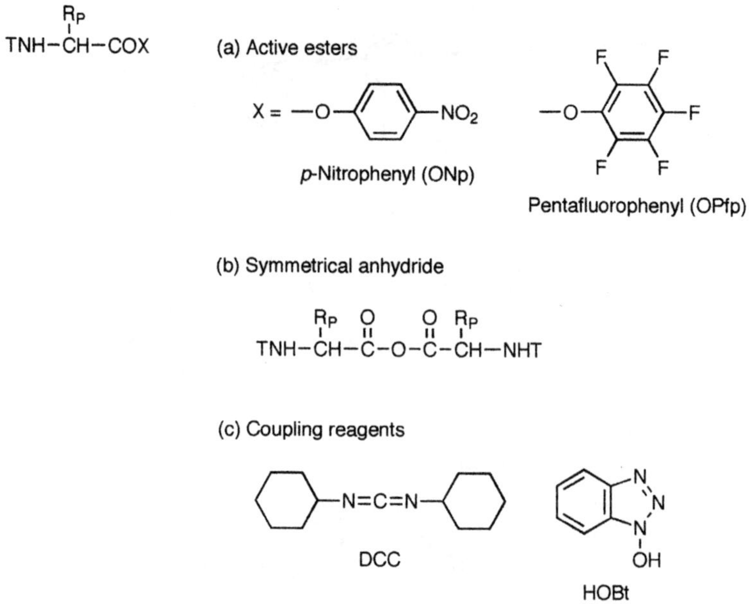

$$\underset{\text{TNH–CH–COX}}{\overset{\text{Rp}}{|}}$$

(a) Active esters

$$X = -O-\!\!\!\!\bigcirc\!\!\!\!-NO_2$$

p-Nitrophenyl (ONp)

Pentafluorophenyl (OPfp)

(b) Symmetrical anhydride

$$\underset{\text{TNH–CH–C–O–C–CH–NHT}}{\overset{\text{Rp O   O Rp}}{|\ \ \ ||\ \ \ \ ||\ \ \ |}}$$

(c) Coupling reagents

$$-N=C=N-$$

DCC

HOBt

Figure 2.2 Activation and coupling methods. (a), (b) Activated N^{α}-protected amino acids; T (temporary protecting group), Boc or Fmoc; R_P, a protected side chain. (c) *In situ* coupling reagent and additive

prior to use from two equivalents of the N^{α}-protected amino acid and one equivalent of the condensing agent dicyclohexylcarbodiimide (DCC), and secondly this is a wasteful process as half of the original (and expensive) carboxylic acid component ends up as carboxylate during the coupling step. This method has generally gone out of favour. The preferred method of coupling in many syntheses today is a one-pot procedure in which DCC is added to a mixture of the carboxyl and amine components. DCC can be used alone or in conjunction with one of the hydroxybenzotriazole (HOBt) family of additives. Activation and coupling proceed concurrently. With the DCC/HOBt combination of reagents, the O-acyl isourea derivative, which forms initially between the carboxylic acid and DCC, reacts with the HOBt to form an active ester. This sequence of events has the advantage in that not only is the lifetime of the racemization-prone O-acyl isourea species reduced, but that the HOBt ester is particularly reactive with amines due to anchimeric assistance (Scheme 2.4).

The du Vigneaud oxytocin synthesis was 'classical' in that at each step of the synthesis the intermediates were isolated, purified and characterized – a time-consuming and often difficult procedure. The invention of the solid-phase approach by Merrifield in 1963 simplified peptide synthesis considerably by allowing the chain to be built up while attached to an insoluble polymeric support. The latter acts, in effect, as an insoluble protecting group for the carboxy terminus. The C-terminal residue of the target peptide is attached to

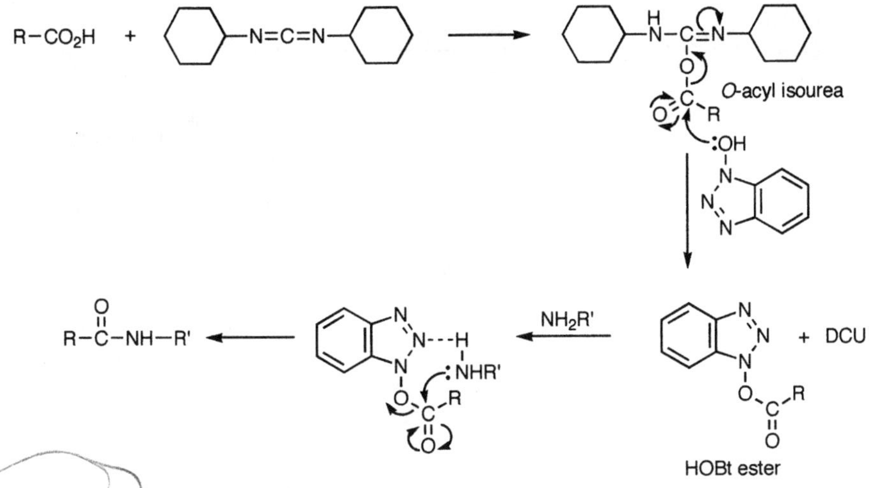

$R-CO_2H$ + [cyclohexyl]$-N=C=N-$[cyclohexyl] $\longrightarrow$

O-acyl isourea

HOBt ester

$R-\overset{O}{\overset{\|}{C}}-NH-R'$ $\longleftarrow$ $\quad$ $\overset{NH_2R'}{\longleftarrow}$ $\quad$ + DCU

Scheme 2.4 DCC/HOBt-mediated amide bond formation; DCU, dicyclohexylurea ($C_6H_{11}NHCONHC_6H_{11}$)

the insoluble support by a benzylic ester linkage through reaction of the N^α-Boc protected amino acid caesium salt **2.5** with chloromethyl-functionalized polystyrene **2.6** in the presence of potassium iodide (Scheme 2.5). The chain is extended by repeated cycles of deprotection of the N^α-Boc group with TFA followed by neutralization with triethylamine to generate the free amine group and then coupling of the next amino acid. At the end of the synthesis, cleavage of the benzyl-type ester linkage, and generation of a 'free' peptide, is accomplished at the same time as removing all of the benzyl-based side-chain protecting groups with liquid HF. As the growing peptide chain remains attached to the insoluble carrier throughout all of the synthetic steps, it can be separated from surplus soluble reagents and solvents by simple filtration and washing procedures. This has the dual advantage of speed, i.e. excess reagents can be used to drive the reactions to completion, and simplicity, i.e. the procedure can be automated. However, a major drawback of the technique is that all purification is left until the chain assembly is complete and the peptide is cleaved from the support. A reasonable quality of crude peptide can generally be obtained if all the coupling steps are made to occur in nearly 100% yield by using excess acylating agent and repeating the coupling steps as necessary. Purification of the peptides is carried out in free solution by a combination of high-performance liquid chromatography, gel permeation chromatography and ion exchange chromatography. Peptide characterization is through amino acid and mass spectral analyses.

Solid-phase chemistry has been developed to such a fine art that for many peptides up to approximately 30 residues in length, chain assembly and cleavage from the resin are, with the appropriate equipment, quite literally

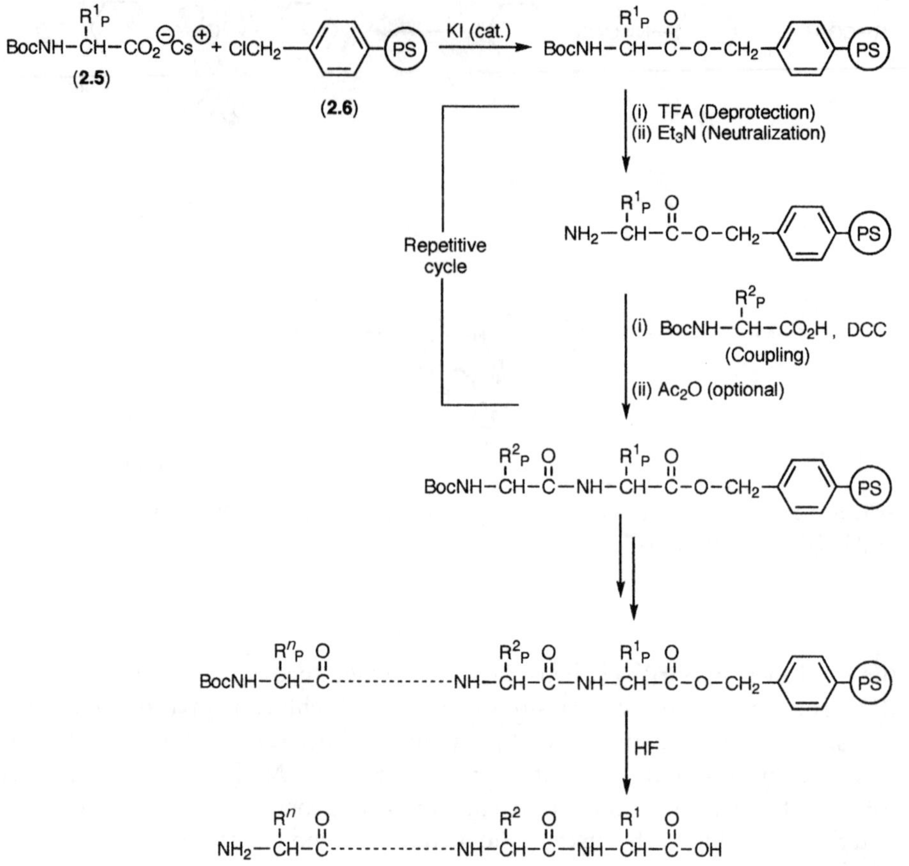

BocNH–CH(R^1_P)–CO$_2^-$Cs$^+$ + ClCH$_2$–⟨C$_6$H$_4$⟩–(PS) $\xrightarrow{\text{KI (cat.)}}$ BocNH–CH(R^1_P)–C(=O)–O–CH$_2$–⟨C$_6$H$_4$⟩–(PS)

(2.5)   (2.6)

(i) TFA (Deprotection)
(ii) Et$_3$N (Neutralization)

NH$_2$–CH(R^1_P)–C(=O)–O–CH$_2$–⟨C$_6$H$_4$⟩–(PS)

(i) BocNH–CH(R^2_P)–CO$_2$H, DCC (Coupling)
(ii) Ac$_2$O (optional)

Repetitive cycle

BocNH–CH(R^2_P)–C(=O)–NH–CH(R^1_P)–C(=O)–O–CH$_2$–⟨C$_6$H$_4$⟩–(PS)

BocNH–CH(R^n_P)–C- - - - - - - - - - - -NH–CH(R^2_P)–C(=O)–NH–CH(R^1_P)–C(=O)–O–CH$_2$–⟨C$_6$H$_4$⟩–(PS)

HF

NH$_2$–CH(R^n)–C- - - - - - - - - - - - -NH–CH(R^2)–C(=O)–NH–CH(R^1)–C(=O)–OH

Scheme 2.5 Merrifield's solid-phase peptide synthesis method. The resin support is a polystyrene suspension cross-linked with 1% of 1,3-divinylbenzene. In DMF, the solvent used for synthesis, the dry polystyrene beads swell two- to sixfold in volume and thus the chemistry of solid-phase synthesis takes place within a well-solvated matrix rather than, as the term would suggest, on the surface of the solid phase. The chain is assembled by first attaching the C-terminal amino acid to the support and then adding the subsequent residues using repeated deprotection/coupling cycles. The C-terminal residue is anchored to the resin by reacting an N^α-protected amino acid, as its caesium salt, directly with a chloromethyl resin to provide a polymer-bound benzyl ester. The next stage of solid-phase synthesis is systematic elaboration of the growing peptide chain. One cycle of chain extension involves three stages, namely deprotection, neutralization and coupling. The Boc group is quantitatively removed by treatment with TFA followed by neutralization of the amine salt with a tertiary amine to liberate the N^α-amine of the peptide resin. The next incoming protected amino acid is attached to the growing peptide chain by a suitable coupling procedure, generally using a two-to fourfold excess of the amino acid to ensure >99% coupling. Coupling is repeated if less than quantitative coupling has occurred, or alternatively the unreacted amino group is permanently acylated (generally with acetic anhydride) to ensure that the chain does not grow during the subsequent synthetic steps.

automated processes. Manual intervention is only necessary during purification and characterization of the peptide.

Direct anchoring of the C-terminal amino acid to the resin has waned in popularity and has been replaced by linkage agents. The latter are bifunctional spacer molecules that on one end incorporate features of a smoothly cleavable protecting group and on the other end allow coupling to a previously fuctionalized support. The linkers serve to link the first amino acid to the resin in two discrete steps and thereby afford better control over this key synthetic step (Scheme 2.6). Furthermore the use of linkers in synthesis offers flexibility as the basic structure, e.g. **2.7**, can be varied; compound **2.8** allows the use of milder reagents to cleave the assembled peptide from the support, while **2.9** enables the preparation of sequences with a modified C-terminus such as a peptide amide.

Together with Merrifield's original pioneering Boc/benzyl combination of protecting groups, the pairing of Fmoc for temporary N^α-protection and t-butyl-based groups for side-chain protection developed by Sheppard dominate solid-phase protocols (Table 2.3). The Sheppard combination is chemically less complex than the Boc procedure (no neutralization is required), uses milder reagents (no HF necessary) and has become the method of choice in many laboratories for the solid-phase synthesis of peptides (Scheme 2.7).

A solid-phase synthesis of oxytocin is outlined in Scheme 2.8. The peptide was assembled using the Fmoc/t-butyl strategy with p-[(α-Fmoc-amino)-2,4-dimethoxybenzyl]phenoxyacetic acid as the linkage agent. This linker yields the peptide amide directly when the protected peptide resin is treated with TFA. The thiol groups of the two cysteines were protected with trityl (Trt) groups. There are a number of groups that are compatible with Fmoc/t-butyl chemistry which provide effective protection for cysteine. Trityl groups are used extensively as the free thiol is generated upon deprotection with TFA. The acetamidomethyl (Acm) group is stable to TFA but can be removed with mercuric acetate or iodine and is the preferred method of protection when a

---

This gives small amounts of truncated sequences which are generally easier to separate from the final product than deletion sequences missing only one amino acid. The coupling step is monitored by carrying out the ninhydrin test on an aliquot of resin to ensure completion – a positive colourimetric response indicates the presence of unreacted N^α-amino groups. At every stage, after attachment of the first residue, and after each deprotection, neutralization and coupling step of the addition cycle, the insoluble polymer–peptide conjugate is washed exhaustively to remove excess reagents and co-products. From the attachment of the first residue to the addition of the last, the peptide–resin conjugate remains in the same vessel, retained there by a sintered glass partition, through which all the washings pass to waste. Once chain assembly has been accomplished the polymer-bound protected polypeptide is treated, in a special polytetrafluoroethylene-lined apparatus, with HF to simultaneously remove all of the protecting groups and cleave the benzyl ester polymer–peptide link giving the crude peptide (R_P^1, R_P^2, R_P^n, permanently protected side-chain functionality of amino acids; PS, polystyrene)

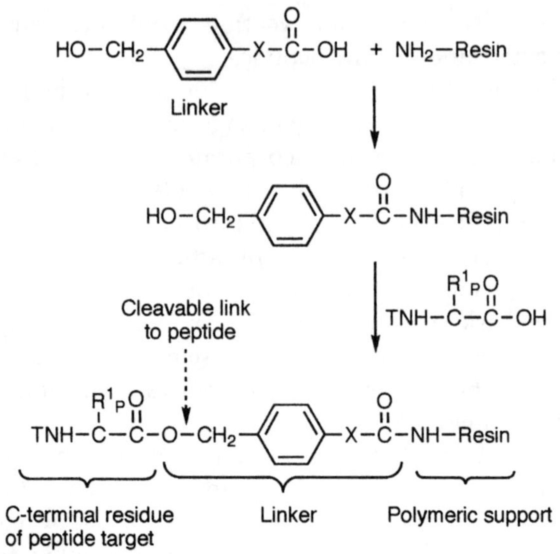

$$HO-CH_2-\text{[benzene ring]}-X-\overset{\overset{O}{\|}}{C}-OH \quad + \quad NH_2-Resin$$

Linker

$$HO-CH_2-\text{[benzene ring]}-X-\overset{\overset{O}{\|}}{C}-NH-Resin$$

Cleavable link
to peptide

$$TNH-\overset{\overset{R^1_P}{|}}{C}-\overset{\overset{O}{\|}}{C}-OH$$

$$TNH-\overset{\overset{R^1_P}{|}}{C}-\overset{\overset{O}{\|}}{C}-O-CH_2-\text{[benzene ring]}-X-\overset{\overset{O}{\|}}{C}-NH-Resin$$

C-terminal residue   Linker   Polymeric support
of peptide target

	N^α protecting group (T)	Cleavage conditions	Resulting C-terminus
(2.7) X = CH$_2$	Boc	HF	Acid
(2.8) X = O−CH$_2$	Fmoc	TFA	Acid
(2.9) X = ----	Fmoc	NH$_3$	Amide

Scheme 2.6   The use of linkers in solid-phase synthesis. In the first step, the carboxyl group of the linker is activated and coupled to an amino-functionalized resin. In the second step of the procedure, the exposed hydroxyl group of the resin-bound linker forms a benzyl ester linkage to the C-terminal residue of the target peptide (R^1_P, protected side-chain of the C-terminal amino acid)

Table 2.3   Protecting groups used in routine solid-
phase peptide synthesis

N^α-Fmoc chemistry	N^α-Boc chemistry
Arg(Pmc)	Arg(Mts)
Asp(OBu^t)	Asp(OBzl)
Cys(Acm or Trt)	Cys(Acm)
Glu(OBu^t)	Glu(OBzl)
His(Boc or Trt)	His(Bom)
Lys(Boc)	Lys(Z)
Ser(Bu^t)	Ser(Bzl)
Thr(Bu^t)	Thr(Bzl)
Trp(Boc)	Trp(For)
Tyr(Bu^t)	Tyr(Bzl)

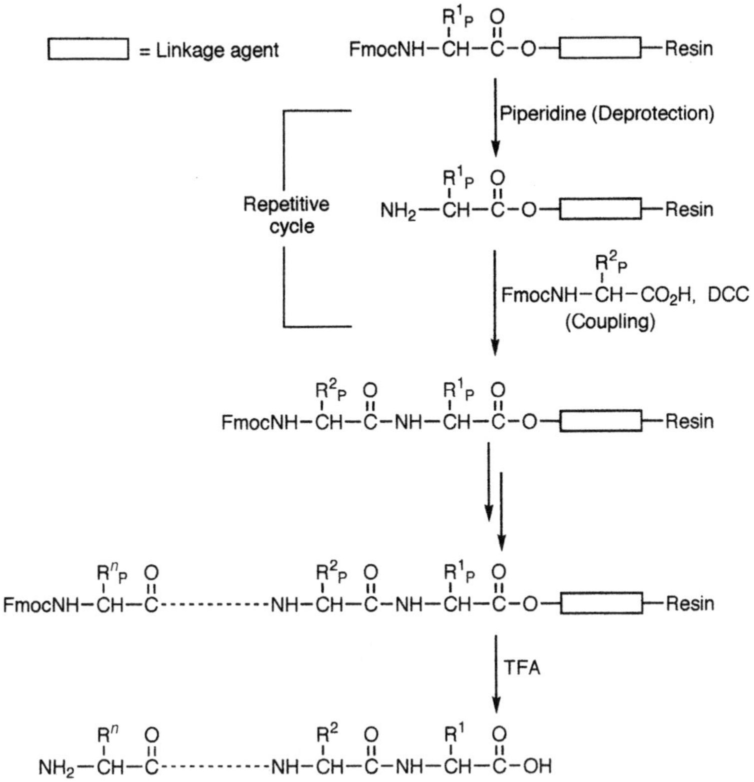

Scheme 2.7 Sheppard's solid-phase assembly of peptide chains by a repetitive two-step cycle of deprotection and coupling. The features of the Sheppard technique which make this a gentler and more popular method of synthesis than the Merrifield approach are first that the temporary N^α-Fmoc protecting group is removed with 20% piperidine in DMF, and secondly that cleavage from the resin and global side-chain deprotection is effected with 95% TFA. No HF is used. The solid support is generally either a cross-linked acrylamide polymer contained within the pores of a rigid matrix or poly(ethylene glycol)–polystyrene materials. These supports can be packed into columns and can be used in continuous-flow as well as in batchwise syntheses. The two techniques differ principally in the method employed for washing the resin between the various synthetic steps. In continuous-flow synthesis, washing is achieved by pumping solvent through the resin bed. In the batchwise process, the peptidyl resin is contained within a fritted reaction vessel and reagents are added in portions through the top of the vessel and removed by either the application of positive nitrogen pressure or vacuum. The Sheppard group introduced the linker approach which has subsequently been incorporated into Merrifield protocols. Linkers generally have carboxyl groups which are coupled, through the formation of an amide bond, on to supports which have been functionalized with amino groups. The C-terminal amino acid of the target peptide is then attached to the exposed hydroxyl groups on the linker. Coupling is carried out with pre-formed active esters or by using activation reagents that generate *in situ* benzotriazolyl esters (R_P^1, R_P^2, R_P^n, permanently protected side-chain functionality of amino acids)

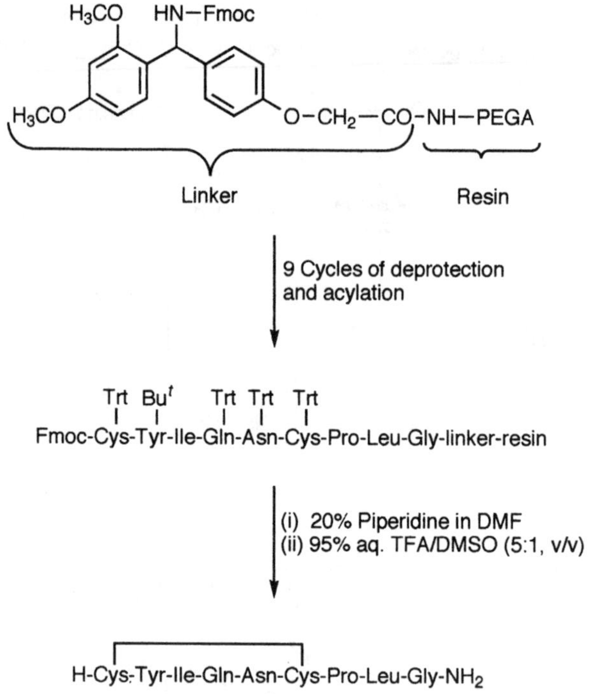

H₃CO   HN—Fmoc

(Linker structure)

H₃CO ... O—CH₂—CO—NH—PEGA

Linker   Resin

9 Cycles of deprotection
and acylation

Trt Buᵗ   Trt Trt Trt
 | | | | |
Fmoc-Cys-Tyr-Ile-Gln-Asn-Cys-Pro-Leu-Gly-linker-resin

(i) 20% Piperidine in DMF
(ii) 95% aq. TFA/DMSO (5:1, v/v)

H-Cys-Tyr-Ile-Gln-Asn-Cys-Pro-Leu-Gly-NH₂

(2.10) Oxytocin

Scheme 2.8 Solid-phase synthesis of oxytocin. The side-chain amino acid protecting groups were Buᵗ for Tyr and Trt for Asn, Gln and Cys. The fully protected N^α-Fmoc amino acid-OPfp esters were coupled in DMF in the presence of a catalyst (3,4-dihydro-3-hydroxy-4-oxo-1,2,3-benzotriazine) (PEGA, poly(ethylene glycol)–polyacrylamide copolymer)

group on sulphur needs to be maintained after cleavage of the peptide from the solid support. Selective disulphide formation in polycysteinyl peptides is therefore possible using Cys(Trt) in combination with, for example, Cys(Acm). In the oxytocin synthesis, the Cys(Trt) peptide was directly oxidized to the disulphide by using TFA in dimethylsulphoxide.

Solid-phase techniques are now routinely used to prepare small quantities (10 to 100 mg) of peptides between five and fifty amino acids in length. Larger quantities of peptides, such as in industrial syntheses, are more commonly prepared by using solution chemistry. Solid-phase principles have been successfully extended to oligonucleotide synthesis (see Chapter 5) and to other fields of synthetic chemistry.

2.4   Gene Cloning

The availability of therapeutic proteins from natural sources is limited by tissue availability, the level at which the protein is produced in such tissues and the possible risk of disease transmission through infected raw materials. These difficulties can be overcome through rDNA technology and many therapeutic proteins, for example, insulin, human growth hormone and interferons, are now produced by such methods.

Every protein, regardless of its source, is produced as a result of expression of a specific gene coding for it. By integrating the DNA for a specific human protein into fast-replicating carrier DNA which can be amplified in bacteria, it is possible to produce enormous numbers of copies of the human gene. Expression of the genes within the chimeric or recombinant DNA yields large amounts of the desired human protein. This is the basis of gene cloning – one of a set of techniques collectively known as recombinant DNA technology.

There are four essential steps to rDNA-derived proteins (Scheme 2.9). First, the DNA encoding the therapeutic protein is generated and then this fragment is incorporated into a carrier or vector. Next, the vector is introduced into a

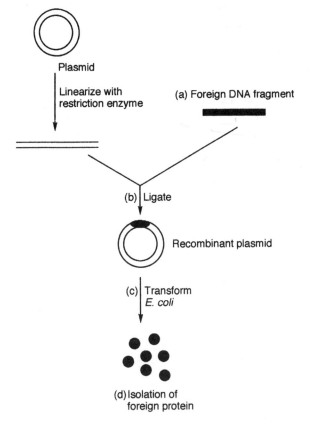

Scheme 2.9 Basic protocol for the expression of foreign DNA in bacteria: (a) the foreign DNA fragment is either synthesized or isolated from genomic DNA and (b) joined to an carrier molecule *in vitro*; (c) the recombinant molecules are transferred to an appropriate host cell where they can replicate; (d) the cells are induced to produce proteins that they would otherwise not synthesize

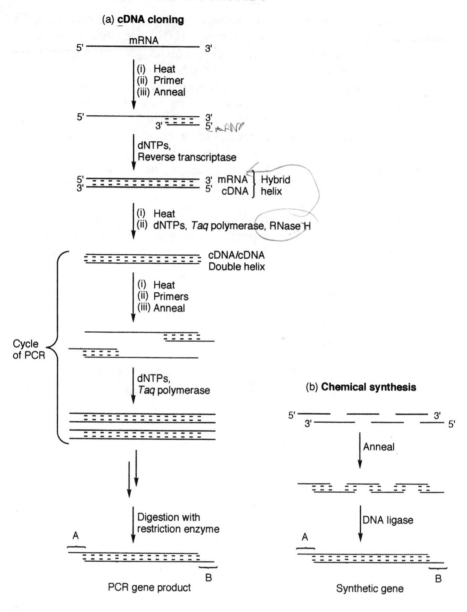

(a) cDNA cloning

(b) Chemical synthesis

host organism which is grown in culture, and finally the desired protein is isolated and purified.

(1)   Obtaining DNA

The DNA to be cloned can be derived directly from natural sources (genomic cloning) or it can be made by copying mRNA (complementary DNA (cDNA) cloning), or it can be chemically synthesized. In genomic cloning, the entire

Scheme 2.10 Synthesis of DNA for cloning. (a) cDNA cloning and amplification using PCR. The mRNA is first denatured by heat and then an oligonucleotide primer complementary to the 3′ terminus of the mRNA is added. By lowering the temperature, base-pairing between the primer and the mRNA is established (annealing). Reverse transcriptase, an RNA-dependent DNA polymerase, in the presence of the four dNTPs, then extends the primer in the 3′ direction thus resulting in the synthesis of a cDNA strand. The hybrid duplex is then denatured by heat and RNase H is then used to hydrolyse the RNA strand. As a result, 3′-hydroxyl priming groups are created which are used by DNA polymerase from *Thermus aquaticus* (a heat-stable polymerase) to initiate the synthesis of a second strand of cDNA. Once double-stranded DNA has been synthesized, it is amplified by using the standard cycle of PCR: (i) denaturation of the double-stranded DNA by heating; (ii) annealing of primers to target sequence; (iii) extension of primers with *Taq* polymerase. Each of the new strands synthesized will act as templates and so there will be an exponential increase in the amount of DNA produced. The cycle is repeated until sufficient DNA is obtained for ligation to the vector. By using primers containing appended restriction sites in the amplification process the DNA product can be digested with the appropriate enzyme to generate 'sticky ends' (A and B), i.e. single-stranded sequences which facilitate the insertion of DNA into the cloning vector. (b) Gene assembly by enzymatically linking together smaller synthetic DNA fragments

genome of a cell is cleaved with restriction endonucleases, enzymes which act like scissors and hydrolyse the phosphodiester bonds of the DNA at specific sites. An enormous number of nucleotide segments (a genomic library) are produced as a result and the DNA fragment encoding the protein of interest must be isolated from the mixture. Making the gene from purified cellular mRNA is mostly done through a combination of reverse transcription of the mRNA and amplification of the resulting cDNA copy by using the polymerase chain reaction (PCR) (Scheme 2.10(a)). The PCR is essentially a method of synthesizing DNA *in vitro*. This method works through the presence of two chemically synthesized oligonucleotides (primers) of around 20 base pairs which are complementary to sequences on either side of the target sequence. Heating the target DNA separates the two strands and then cooling the single-stranded templates in the presence of the primers causes the oligonucleotides to base-pair and form short double-stranded sections. Each of these sections can be extended by DNA polymerase in the presence of the four deoxynucleotide triphosphates (dNTPs) to make a copy of the original sequence. As nucleotides are added at the 3′ end of the molecule, the two chains will grow in opposite directions and the result will be two complete double-stranded copies of the target. Each new cycle of PCR (i.e. heating–cooling/annealing–polymerization) gives rise to twice the number of copies of template DNA from the previous cycle and thus the original DNA is amplified exponentially by approximately 2^n-fold, where n is the number of cycles. By using primers which incorporate sequences that specify a recognition site for a restriction enzyme in the amplification process, the DNA can be digested to generate single-stranded termini ('sticky ends') which are compatible for joining to a vector which has been treated with the same restriction enzyme (see below).

Genes can also be constructed from multiple segments of chemically synthesized oligonucleotides (Scheme 2.10(b)). These short fragments hydrogen-bond with each other in an unambiguous way and are then covalently joined with DNA ligase. As with the cDNA method, the chemically synthesized gene is designed to contain restriction enzyme recognition sites.

(2) Constructing rDNA

The next step is to insert the DNA to be cloned into the vector (or cloning vehicle). The most commonly used vector is a plasmid, namely a circular DNA molecule that occurs naturally in bacteria and which replicates independently of the chromosomes. A number of plasmids have been artificially modified and constructed as cloning vectors, e.g. pBR322 (Figure 2.3).

Desirable features of a cloning vector are as follows:

(1) A bacterial origin of replication. This serves as a start signal for DNA polymerase and ensures that plasmid DNA molecules will be replicated by the host cell.

(2) Genes that encode antibiotic resistance. These genes are capable of rendering a bacterium tolerant to antibiotics which is extremely useful as it allows bacteria that have taken up the plasmid to be identified and selected (Scheme 2.11). If cells are plated on to a medium containing an appropriate antibiotic, e.g. ampicillin, only those which contain plasmid will grow to form colonies. The presence of a second antibiotic resistance gene allows for the selection of those plasmids which contain inserted DNA; inserting foreign DNA into a restriction site within an antibiotic-resistant gene such as tetracycline inactivates this selectable marker.

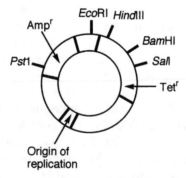

Figure 2.3   Schematic representation of vector pBR322 (termed after its developers Bolivar and Rodriguez). The vector contains an origin of replication, ampicillin (Amp^r) and tetracycline (Tet^r) selectable marker genes and a number of restriction sites; those which are particularly useful for gene cloning are shown. The complete DNA sequence of pBR322 consists of 4361 base pairs

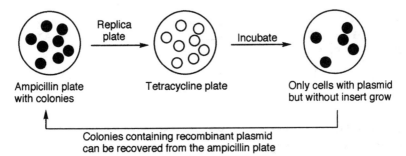

Scheme 2.11 Screening of recombinant plasmids carrying two drug-resistant markers. A replica plate is made by pressing a velvet pad over the surface of the plate and lifting it off and then pressing in on to a fresh plate, thus inoculating it with cells in a pattern identical to that of the original colonies. Plasmids with both functional Tet[r] and Amp[r] genes will grow on the tetracycline plate, while clones of cells that fail to grow on the second plate, because of loss of antibiotic resistance, must not contain inserts and can be recovered from their corresponding colonies on the first plate. Reproduced from Walker, J.M. and Rapley, R., (eds), *Molecular Biology and Biotechnology, 4th Edition*, 2000, by permission of the Royal Society of Chemistry.

Plasmid colonies which have grown on ampicillin but not on tetracycline therefore contain plasmids with inserts and thus can be differentiated from plasmids which have merely recircularized without an insert.

(3) A variety of cleavage sites for restriction endonucleases (Scheme 2.12). These are used to open or linearize the circular plasmid. Linearizing a plasmid allows a fragment of DNA to be inserted and the circle closed. Ideally, any particular site should be present only once in a cloning vector and in a non-essential region. Using a restriction enzyme that has a staggered cleavage, i.e. the two strands of DNA are not cut opposite each other but are offset by several nucleotides, to cut the plasmid generates a linear molecule with protruding single-stranded termini. If the same restriction enzyme recognition sequence is incorporated into the termini of the DNA which is to be inserted, then when this molecule is digested and added to the linear plasmid the complementary ends of the two DNA segments base-pair and they can then be joined with DNA ligase. This is the actual recombinant part of the cloning procedure (Scheme 2.13).

Genes of eukaryotic organisms, such as humans, differ in a number of fundamental ways from those of bacteria such as *Escherichia coli* (*E. coli*), the host organism most extensively used for gene cloning. These differences mean that in order to achieve efficient expression of human genes in a bacterial cell, certain elements must be supplied by the cloning vehicle. First, the genes of *E. coli*, unlike those of eukaryotic cells, do not have intervening sequences (introns) and so the bacterium has no means of removing such sequences from

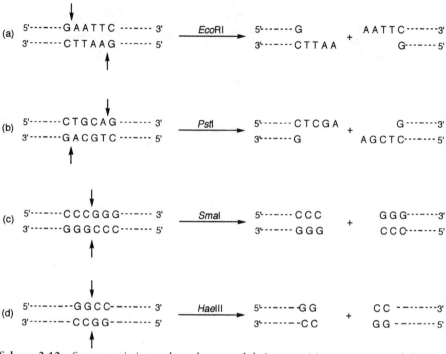

Scheme 2.12   Some restriction endonucleases and their recognition sequences and cleavage sites (indicated by arrows). Cutting can be divided into three types. In some cases ((a) and (b)), the cleavages in the two strands are staggered and because of the symmetry of the recognition sequences this generates mutually cohesive termini ('sticky ends'). In the cases of *Eco*RI, the protruding single-stranded ends have 5'-termini, while other enzymes such as *Pst*I have a staggered cleavage that generates single-stranded 3'-termini. Further enzymes have a flush cleavage and produce blunt-ended fragments, e.g. *Sma*I and *Hae*III in reactions (c) and (d), respectively

primary transcripts of eukaryotic genes. This difficulty can be overcome by using a human gene that is either the cDNA copy of the mature mRNA or is chemically synthesized (see Scheme 2.10). Secondly, in order for a bacterium, i.e. a prokaryotic cell, to transcribe a eukaryotic gene it is necessary to insert a prokaryotic promoter upstream from the human gene. A promoter is a relatively short sequence of DNA where RNA polymerase and associated regulatory proteins bind and begin transcription. There is no generic promoter system that offers optimal high-level expression for every gene. Most plasmid expression vectors use either the promoter of the *lac* (lactose metabolism) or of the *trp* (tryptophan metabolism) operons from *E. coli* or the β-lactamase promoter from the plasmid pBR322. Finally, initiation of translation requires not only a start codon (AUG) but also a ribosome binding site. The ribosome binding site is a nucleotide sequence upstream from the initiation codon of the mRNA transcript which is complementary to, and can therefore base-pair with, RNA in the ribosomes. In order to achieve efficient expression of a

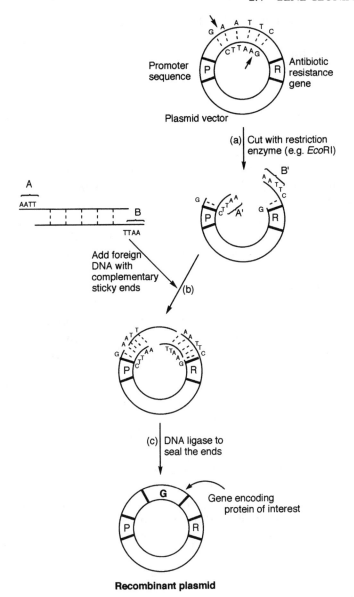

Scheme 2.13 Formation of a recombinant plasmid. (a) Cleavage of plasmid with *Eco*RI to produce a linear DNA molecule with two sticky ends consisting of the single-stranded DNA sequence AATT. (b) These linear molecules are combined with the foreign DNA which has been prepared for cloning and has the same sticky ends as those on the plasmid DNA. (c) The complementary sticky ends of the two DNA molecules anneal and the mixture is treated with DNA ligase to covalently seal the single-stranded breaks and generate the circular recombinant plasmid

eukaryotic gene within *E. coli*, it is necessary to provide that gene with a bacterial ribosome binding site. Expression vectors have been developed which contain appropriate promoter and ribosome binding sites positioned just before one or more restriction sites for the insertion of foreign DNA.

(3)   Cloning

Having inserted a segment of foreign DNA into a suitable vector, the next step is to introduce the vector into a host organism. The organism most frequently used is the bacterium *E. coli* and the recombinant plasmids are introduced into the bacterial cells by transformation. Bacteria do not normally take up DNA from their surroundings but they can be induced to do so by treatment with calcium chloride. The bacteria take up the DNA and as these cells multiply the recombinant plasmids replicate and express their products, including the human protein. However, transcription from the *lac* promoter is negatively regulated by the *lac* repressor, a protein that binds specifically to the promoter sequence. As long as the repressor sits on the DNA, RNA polymerase is prevented from binding and initiating transcription. Expression from the *lac* promoter is stimulated by neutralizing (derepressing) the repressor through the addition of an inducer such as isopropyl β-D-thiogalactoside (IPTG) (Figure 2.4).

The problems of procuring a prokaryotic promoter and ribosome binding site to obtain expression of eukaryotic DNA in *E. coli* can be obviated by constructing a fusion gene. The control region and N-terminal coding sequence of an *E. coli* gene is ligated to the coding sequence of a eukaryotic gene. When introduced and cloned in *E. coli*, RNA polymerase recognizes the promoter as native and transcribes the gene. The 5′-end of the mRNA will also therefore be native and consequently interacts normally with a ribosome to commence protein synthesis. The proteins that result will be chimeric, with the N- and C-terminals being derived from the prokaryotic and eukaryotic genes, respectively (see, for example, the synthesis of insulin in Section 2.6.1 below).

(4)   Isolation

Frequently, human proteins expressed in *E. coli* accumulate in the cell cytoplasm in the form of inclusion bodies. These are insoluble aggregates of the partially folded protein. In this form, the protein is inactive and must be refolded into the native conformation. Refolding can be a complicated process but in essence it entails adding denaturants, such as guanidinium hydrochloride, to solubilize the aggregated polypeptide and then finding the right conditions to promote refolding into the biologically active conformation. Suitable conditions required for such renaturation can vary from protein to protein and it is not always possible to obtain high yields of activity. To avoid

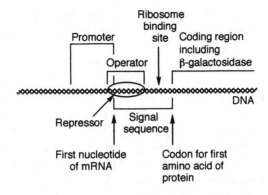

Figure 2.4   The *lac* promoter-operator system. The *lac* promoter is responsible for the expression of several genes involved in the transport and metabolism of lactose. One of these genes encodes β-galactosidase (β-gal), an enzyme which catalyses the cleavage of lactose to galactose plus glucose. The expression of β-gal is under the control of a repressor which is sensitive to the levels of the available lactose. In the absence of lactose (the inducer), the *lac* repressor binds tightly to the *lac* operator, thereby blocking the progress of the RNA polymerase and consequently transcription of the *lac* genes and synthesis of β-gal is inhibited. This strategy ensures efficiency in the cell by preventing the expression of genes when there is no substrate for the protein product to act upon. When lactose is present, it binds to the repressor. This alters the repressor's three-dimensional structure which causes it, the repressor, to loosen its hold on the DNA so that it no longer binds to the operator. The operator is thus activated (derepressed), the β-gal gene is turned on, β-gal is synthesized and this then breaks down the lactose to galactose and glucose. An artificial inducer such as IPTG can derepress the operator and thereby stimulate gene expression. The advantage of using IPTG is that it is not degraded by β-gal and is taken up by the cells more readily than lactose. This system is referred to as negative control because the regulatory protein, the *lac* repressor, prevents gene expression. Reproduced from Drlica, K., *Understanding DNA and Gene Cloning* 1984 © John Wiley & Sons, Inc. Reprinted by permission of John Wiley & Sons, Inc.

formation of inclusion bodies it is possible to direct protein secretion from the cytoplasm of *E. coli* into the periplasm, i.e. the space between the inner and outer membranes of the bacterial cell, or even into the supernatant. Proteins are transported through the cell membrane by hydrophobic N-terminal signal sequences. Directing synthesis to the periplasm can be achieved by including genes which encode signal sequences into the expression vector, although the presence of the signal peptide does not always ensure efficient protein translocation through the inner membrane. The oxidizing environment of the periplasm facilitates proper protein folding and increases the likelihood that the product can be recovered in an active form. Furthermore, *in vivo* cleaving of the signal peptide during translocation to the periplasm is more likely to yield the authentic N-terminus of the target protein (see also below).

Foreign proteins can be partially degraded in *E. coli* by proteases. Such degradation products are potentially immunogenic and can present a problem in the production of therapeutics. Thus, isolated protein, whether refolded from inclusion bodies, recovered from the periplasm or extracted from the supernatant, is subjected to additional purification by using a range of chromatographic techniques.

Bacteria do not process proteins in exactly the same way as do mammalian cells so that the expression of human proteins in bacteria does not always yield the active product or any products at all. Many human proteins must be post-translationally modified in order to exert their biological activity and post-translational modifications such as glycosylation and phosphorylation are not observed in bacteria. All proteins produced in bacteria have N-formyl methionine as their N-terminal residue. Although bacteria have the enzymic capability to deformylate and remove this residue, eukaryotic proteins produced in bacteria may have a methionine at their N-terminus which can alter their biochemical characteristics. Methionine can, however, be removed from E. coli protein products by the action of cyanogen bromide (Scheme 2.14). This reagent causes peptide bond cleavage on the carboxy side of methionine residues with concomitant conversion of the methionine to homoserine lactone which is in equilibrium with homoserine. This process is, of course, only applicable to the production of eukaryotic proteins not containing internal methionines.

In order to obtain properly modified and folded proteins exhibiting the desired biological activity, it is frequently necessary to express genes in eukaryotic expression systems. Yeasts such as *Saccharomyces cerevisiae* are capable of glycosylating proteins but the level and type of modification is limited. Only mammalian cell lines possess all of the biochemical and cellular

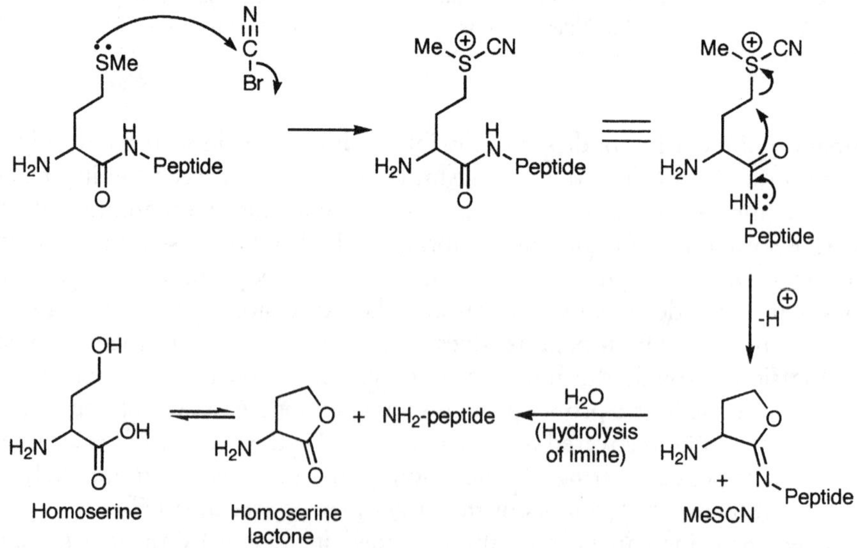

Scheme 2.14   Cleavage of peptide chains C-terminal to methionine residues with cyanogen bromide

machinery necessary for a full range of protein modifications. Mammalian cell expression is, however, less efficient for protein production than *E. coli* expression systems and typically gives relatively little product (less than 50 mg per litre) compared with the bacterial route (many grams of protein per litre of *E. coli* culture).

2.4.1 Growth hormone

Growth hormone is used to treat children with growth retardation. Until recombinant methods were developed the only source of the protein was from pituitary glands obtained at post-mortem. Human growth hormone (hGH) is 191 amino acids in length, which in turn, is too large to be chemically synthesized but it can be made by cloning cDNA (Scheme 2.15). The cDNA, prepared from mRNA extracted from pituitary tissue, carries a nucleotide sequence coding for a (mammalian) secretion signal. This sequence must be removed since expression in bacteria would otherwise afford the prohormone. The cDNA has a *Hae*III restriction site in the sequence coding for amino acids 23 and 24 and an *Sma*1 site located in the untranslated sequences at the 3′ end of the gene. Treatment of the cDNA with *Hae*III and with *Sma*1 gives a 512 base-pair sequence coding for amino acids 24–191 of hGH. As cleavage with *Hae*III removes not only the signal sequence but also the sequence coding for the first 23 amino acids of the hGH gene, it is necessary to replace this latter fragment with synthetic DNA. A chemically synthesized 84 base-pair sequence containing codons for the N-terminal region, an ATG (i.e. methionine) translation initiation site and *Eco*RI and *Hae*III sticky ends is then enzymatically ligated to the 512 base-pair fragment to create the complete 1–191 sequence. The resulting construct is inserted downstream from two *lac* promoters in tandem and the recombinant plasmid cloned in *E. coli*. Met-hGH is produced as a soluble protein (approximately 300 mg per litre of culture) which can be readily purified.

The presence of the additional methionine does not affect the biological properties of GH although low levels of non-neutralizing antibodies have been formed in some patients receiving prolonged treatment with Met-hGH. In the synthesis of hGH in the pituitary gland, the N-terminal methionine is part of a secretory signal sequence which is enzymically cleaved *in vivo* thereby removing the methionine and generating the 191 amino acid sequence. Furthermore, hGH identical to native protein can be produced in *E. coli* by cloning the coding sequence next to a bacterial signal sequence. The signal sequence specifies secretion of the protein to the periplasmic space. A periplasmic protease cleaves the signal sequence including the methionine, thus leaving the native hGH which can be extracted. Both methioninated and non-methioninated recombinant growth hormone are in clinical use.

Therapeutic proteins produced in *E. coli* include insulin, growth hormone and the interferons while many higher-molecular-weight proteins such as

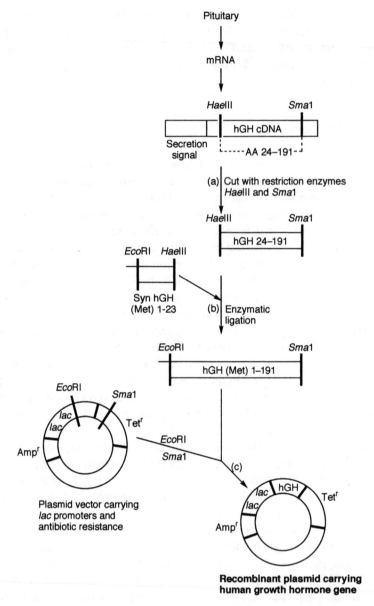

Scheme 2.15   Construction of a recombinant plasmid that directs the bacterial synthesis of human growth hormone. (a) Restriction enzyme digestion of cDNA, synthesized from the pre-hormone mRNA, to remove the mammalian N-terminal signal sequence as this would not operate in *E coli*. (b) Ligation of the cDNA fragment with a chemically synthesized DNA segment corresponding to the N-terminal region. (c) Insertion of the fused gene *via* *Eco*RI and *Sma*I sites into a plasmid vector containing two copies of the *lac* promoter (Amp^r, ampicillin resistance gene; Tet^r, tetracycline resistance gene)

antibodies and factor VIII are more successfully produced in mammalian cell culture systems. The list of human diseases that can be successfully treated by using recombinant products continues to grow. The identification of novel genes by the Human Genome Project is likely to reveal more potentially useful protein products that can reasonably be developed.

2.5   Transgenic Procedures

Transgenic technology is a relatively new technique which aims to use animals as bioreactors for the production of large amounts of therapeutic proteins. As with expression systems based on cultured mammalian cells, the mammary gland of an animal is capable of performing post-translational modifications vital to the activity of certain pharmacologically active human proteins. In transgenic methods, the expression of a foreign gene is targeted to the mammary gland of an animal so that the foreign protein is secreted directly into its milk. Not only can the protein be easily harvested but, as it is physically removed from the animal's circulatory system, it prevents any physiological problems in the animal associated with high levels of protein. The relatively low operating costs and the potentially unlimited expansion of the producer animals through breeding programmes make this production route to glycosylated proteins an appealing alternative to cell culture systems.

The procedure involves fusing the foreign protein gene (either cDNA or genomic DNA) to the promoter sequence of the gene for a milk protein such as β-lactoglobulin or β-casein and then the fusion construct is micro-injected into an ovum which is subsequently fertilized and implanted into a surrogate mother. If the foreign DNA is successfully incorporated into the chromosomes of the cell nucleus, then the transgenic animal, commonly a sheep or goat, is capable of producing the protein of interest. At birth, the offspring are screened for the presence of the transgene but expression levels cannot be analysed until the first lactation resulting from pregnancy. Thus, transgenic females are brought to maturity and mated and on the birth of their offspring the founder female with the highest expression levels of foreign protein in her milk can be identified. The time required for the first lactation may be as much as 18 months depending on the targeted livestock species. This is one obvious drawback to the approach which is compounded if the transgenic animal is a male.

Following first milking, the best yielding female is identified and used to generate a male which becomes the founder member of the production flock. Semen from the founder male is used to generate the production flock so that all producer animals (second-generation females) have the same father in order to minimize genetic variation. However, there is no guarantee that the expression levels of the founder female will be transmitted to subsequent transgenic offspring. Stable transgene transmission is thus a second difficulty with the technique. Furthermore, it takes a number of years to expand from the founder animal to a flock or herd of useful size.

Subtle differences in the way in which post-translational modifications are carried out in animals and humans may result in a glycosylation variant of the human protein being secreted into the transgenic animal's milk. Most slight alterations to a protein's glycosylation patterns do not affect its biological activity. Some variants may indeed be beneficial which could actually improve the way the proteins function as therapeutic drugs.

Tissue plasminogen activator (tPA), a thrombolytic agent (see below) and α_1-antitrypsin are therapeutic proteins which have been successfully produced in transgenic goats and sheep, respectively. α_1-Antitrypsin is a 394 amino acid glycoprotein normally present at 2 g per litre in plasma. It serves as a potent inhibitor of the protease elastase and as such prevents damage to lung tissue by neutrophil elastase. Genetic deficiencies resulting in low circulating concentrations of α_1-antitrypsin is one of the most common hereditary disorders to affect persons of northern European descent. This condition, known as emphysema, which can also be brought about by smoking, is often life-threatening and sufferers require regular infusions of α_1-antitrypsin to help them breathe more easily. Approximately 200 g of the glycoprotein per patient per year is required and it can be obtained from the pooled plasma fraction of donated blood. α_1-Antitrypsin levels of approximately 35 g per litre have been observed in individual transgenic sheep and so a single animal could produce 10 kg of protein in a single lactation period – enough to treat 50 patients for a year.

Cloning of a sheep from an adult cell has also been demonstrated and 'Dolly', the first such animal, was born in 1997. Dolly has subsequently given birth to a healthy lamb, 'Bonnie', through normal mating and gestation. A combination of the transgenic procedure with adult cell cloning has the potential to be a technique which could produce flocks of genetically identical animals each secreting a human protein.

2.6   Examples of Direct Replacement Strategy

2.6.1   Insulin

Insulin (2.10) is a 51 amino acid protein which plays a central role in the metabolism of carbohydrates. The level of insulin is determined by blood glucose levels; an increase in the concentration of blood glucose induces the secretion of insulin from the pancreas which, in turn, promotes the uptake of glucose by tissues, particularly the liver and muscles. Blood glucose levels return to normal values and this in turn decreases the rate of insulin release. Failure to produce insulin results in a high concentration of blood glucose and the condition of diabetes mellitus, more commonly just known as diabetes.

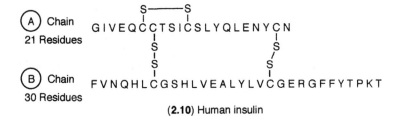

(2.10) Human insulin

There are two types of diabetes – type I, or insulin-dependent diabetes, and type 2, non insulin-dependent diabetes or maturity-onset diabetes. In people with type I diabetes, the pancreas produces little or no insulin and treatment means daily injections of insulin combined with a healthy balanced diet. People with type 2 diabetes produce some insulin, although not enough for the body's needs, but are generally able to control their glucose levels through diet alone.

Traditionally, the supply of insulin for the treatment of diabetes type I has been from the pancreatic tissue of slaughterhouse pigs and cattle, although the insulin from these species is not identical to the human protein. Insulin is composed of two polypeptide chains linked together by two disulphide bonds. The A-chain is composed of 21 amino acid residues while the B-chain contains 30. Human insulin differs in sequence from porcine insulin by one amino acid, i.e. the carboxyl terminal residue of the B-chain. Threonine forms the carboxyl terminus of the human B-chain whereas an alanine residue occupies this position in the porcine protein. In bovine insulin, there is also a C-terminal alanine in the B-chain with additional sequence differences around the disulphide bond of the A-chain. Alanine and valine residues occupy positions A8 and A10, respectively, in the bovine protein compared with threonine and isoleucine at these positions in human insulin. The similarities in structure and biological activity of the porcine and bovine insulins to human insulin has made it possible for the animal proteins to be used in the treatment of diabetes. However, the human immune system treats the animal insulins as foreign and generates neutralizing antibodies which may make it necessary to administer an increased dosage of the porcine or bovine protein to the diabetic patient in order to achieve the desired biological effect.

Although total chemical synthesis of insulin is possible, the process is not economically viable. The human sequence can, however, be prepared from the porcine protein by using a semi-synthetic approach (Scheme 2.16). The procedure involves replacing the C-terminal alanine in the B-chain of porcine insulin with threonine. Treatment of porcine insulin with the enzyme trypsin hydrolyses the Lys^{29}–Ala^{30} (residues 29 and 30) amide link but, as this enzyme acts at the carboxyl side of basic residues, the bond between amino acids 22 and 23 of the B-chain is also cleaved. Thus, rather than adding a single threonine amino acid to the residual insulin protein a C-terminal octapeptide must be coupled. To prevent unwanted reactions during the coupling step,

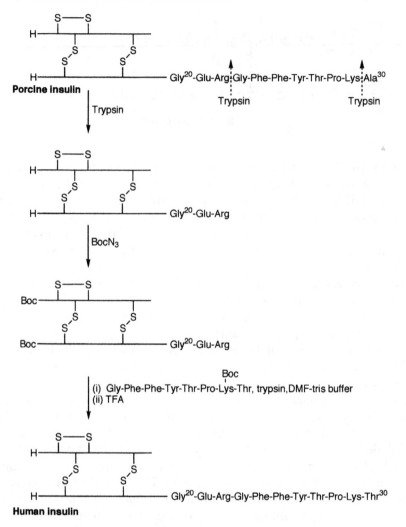

Scheme 2.16 Semi-synthesis of human insulin. Desoctapeptide (B23–B30) insulin, prepared by tryptic digestion of porcine insulin, is coupled to a synthetic octapeptide corresponding to position B23–B30 of human insulin

both fragments are partially protected. The residual insulin fragment has six carboxyl groups and two primary amino groups and these latter functionalities can be effectively masked during the coupling step by acylation with $BocN_3$ (or another Boc-introducing agent). The bis-Boc fragment is then coupled with an excess of the partially protected synthetic human B-chain C-terminal octapeptide. The best results are obtained if the coupling is enzyme-mediated rather than driven by DCC, as the reaction is more specific and free from racemization. In high concentrations of organic co-solvents, trypsin can catalyse formation of amide bonds rather than cause their hydrolysis. In a final step the Boc protecting groups are removed to give the desired human insulin sequence, which is then purified.

Insulin is naturally synthesized as a single peptide, pre-proinsulin, in the pancreas by specialized cells clustered in small regions called the islets of Langerhans (Scheme 2.17). The pre-sequence of pre-proinsulin, a 23 amino acid N-terminal peptide, is responsible for the transport of the molecule from the ribosome, which assembled the chain, across the rough endoplasmic reticulum (RER) membrane into the interior. The N-terminal signal peptide is enzymatically removed from the pre-proinsulin during this process to generate proinsulin. The latter is composed of the amino acid chains that will form insulin and a connecting 30 residue peptide, or C-peptide, that links the end of one chain to the beginning of the other. The connecting link is important in providing the proper alignment of the molecule for formation of the correct disulphide bonds. Proinsulin is folded in the RER, the disulphide bonds are formed and it is then transferred to the Golgi apparatus where it is packaged into granules. Proteolytic conversion of proinsulin to insulin and the C-peptide occurs inside the granules and it is here that the insulin is stored until an extracellular signal, such as an increase in the glucose concentration, in the blood stimulates its secretion.

Bacteria can be made to produce insulin (Scheme 2.18). The procedure entails chemically synthesizing two oligonucleotides which encode the 21 amino acid A chain and 30 amino acid B chain. In addition to the basic 63 and 90 nucleotides coding for the amino acid sequence of the chain, each synthetic gene is extended at its 5' terminus with an ATG (methionine) initiation codon, at its 3' end by a translation termination signal, and has *Eco*RI and *Bam*HI sticky ends. Two expression vectors are constructed, i.e. one for the A chain

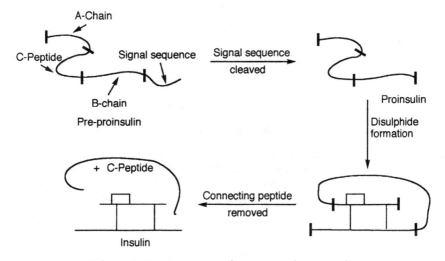

Scheme 2.17   Processing of pre-proinsulin to insulin

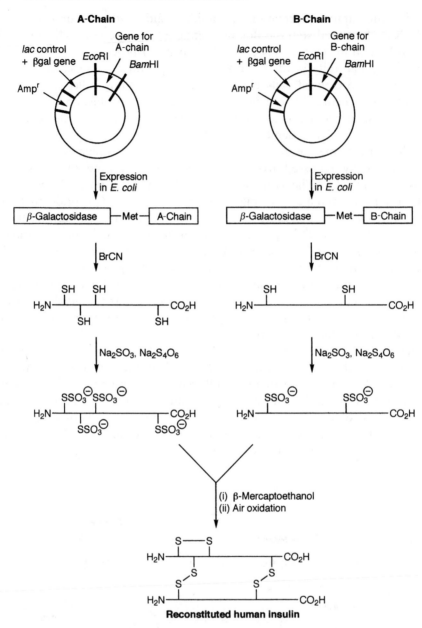

Scheme 2.18   Bacterial synthesis of human insulin. Synthetic genes for the A- and B-chains are spliced into the *Eco*RI site at the 3′ end of the β-galactosidase gene. After expression, the fusion proteins are cleaved with cyanogen bromide and the insulin chains are sulphonated. Subsequent mixing under disulphide interchange conditions affords active insulin

and one for the B-chain. The synthetic genes are separately inserted, along with elements of the *lac* operon, into the plasmid pBR322 at the *Eco*RI site near the C-terminus of the *E. coli* β-galactosidase gene. In bacteria, on induction of β-galactosidase expression, these plasmids direct the synthesis of fusion proteins with β-galactosidase sequences at the N-terminus and insulin A- or B-chain at the C-terminus. The hybrid proteins constitute 20% of the total cellular proteins and precipitate as insoluble aggregates which can be redissolved in guanidinium hydrochloride and formic acid. Insulin does not contain any internal methionine residues and so the A- and B-chains of insulin can be cleaved from the β-galactosidase carrier by exploiting the methionine linker and treating the solubilized hybrid proteins with cyanogen bromide. The A- and B-chains are reconstituted into active insulin by disulphide bridge formation using sodium dithionate and sodium sulphite. Insulin produced in this way is biologically active, although the yield of native insulin from disulphide bond formation is relatively poor.

Insulin obtained by recombination in *E. coli* was licensed for use in 1982 and was the first recombinant protein to be manufactured and brought to the market by using bacteria as a production system. More recent efforts to produce recombinant insulin have used DNA specifying the insulin precursor, proinsulin. The latter is likewise produced as a fusion protein in *E. coli*, but reaction with cyanogen bromide, cysteine *S*-sulphonation, purification and disulphide interchange gives the correctly folded molecule of proinsulin. Conversion of proinsulin to insulin is accomplished by controlled digestion with trypsin and carboxypeptidase B. These conditions remove the C-peptide but do no apparent damage to the insulin part of the molecule in spite of the existence of potential cleavage sites for these enzymes within the insulin structure. An increasing proportion of the world's market for insulin is likely to be met by recombinant human insulin produced in this way.

Proinsulin is 5 to 10% as biologically active as insulin but has a longer plasma half-life; its metabolic clearance rate is 20 to 30% that of insulin. Long-acting insulins such as proinsulin have a role in treating patients with type 2 diabetes who still retain some endogenous insulin secretion but who nevertheless require insulin therapy. Recombinant DNA technology has also enabled the production of other insulin analogues with altered pharmacokinetic profiles. Insulin lispro with a lysine at B28 and a proline at B29 (i.e. the reverse of the natural Pro^{28}, Lys^{29} sequence) is a faster-acting insulin and can be given immediately before meals (5 to 15 minutes) rather than 20 to 40 minutes before as with unmodified insulin.

The nearest thing to a cure for diabetes is a pancreatic transplant but this is rarely undertaken as afterwards it requires a lifetime of immunosupressant drugs to prevent rejection by the body. Transplantation of pancreatic islet cells is an alternative approach which shows promise. Early results indicate that a reasonable degree of glycaemic control is achievable without immunosuppression and the first eight patients to successfully receive pancreatic islet cells have, at the time of writing (May 2001), been living without insulin for about a year.

2.6.2   *The fibrin and fibrinolytic cascades*

Maintenance of normal blood flow requires that the blood should remain fluid at all times. Conversely, staunching of blood loss following injury requires that the blood should also be capable of forming a solid plug rapidly. These conflicting demands are controlled by complex protein–protein interactions of the blood coagulation and fibrinolytic systems (Scheme 2.19). The series of reactions culminating in formation of a blood clot is the coagulation or fibrin cascade. Dissolution of blood clots by degradation of fibrin to soluble fragments is the fibrinolytic system and this cascade is considerably shorter than the coagulation sequence. Both systems are protease cascades, i.e. series of proteolytic enzyme reactions in which each reaction is activated by the previous one and, once initiated, it proceeds to the final one. The power of the cascade system results from cumulative amplification – activation of a single molecule catalyses the activation of many molecules of the next protein in the sequence. This allows a rapid response to a physiological stimulus.

Injury and the resulting tissue damage leads to activation of factor IX, i.e. factor IXa. (*Factor* is the term given to the principal proteins of the cascade and each of these has a Roman numeral designation. Activated forms of the factors are indicated by the addition of the letter 'a' to the factor number.) Factor IXa activates factor X by cleavage of an Arg–Ile bond, while factor Xa converts prothrombin to thrombin which, in turn, hydrolyses specific Arg–Gly bonds of soluble fibrinogen to produce insoluble fibrin. Aggregation of fibrin molecules forms a soft clot which is subsequently converted into a hard clot (thrombus) by the covalent cross-linking of specific amino acid side-chains.

Under normal circumstances, clotting must not occur and a number of endogenous inhibitors maintain blood in a fluid state. A system which is out of balance results in either excessive bleeding or thrombosis and normalization may involve either potentiating or inhibiting coagulation. No one reaction is rate-limiting, and enhancement or inhibition of any individual step in the cascade may lead to enhancement or inhibition of the entire sequence.

Coagulants

Abnormalities resulting from the absence or defective action of any of the blood clotting proteins leads to poor coagulation. The majority of hereditary disorders characterized by poor coagulation ability are due to a deficiency in factors VIII or IX. Factor IX is a central component in the cascade and is activated through proteolysis by factor XI. Factor VIII does not exhibit proteolytic activity but is an accessory protein that participates with protease factor IX in the activation of factor X. A lack of factor VIII thus substantially decreases the rate of formation of factor X and gives rise to the clinical condition of haemophilia A, while a deficiency in factor IX leads

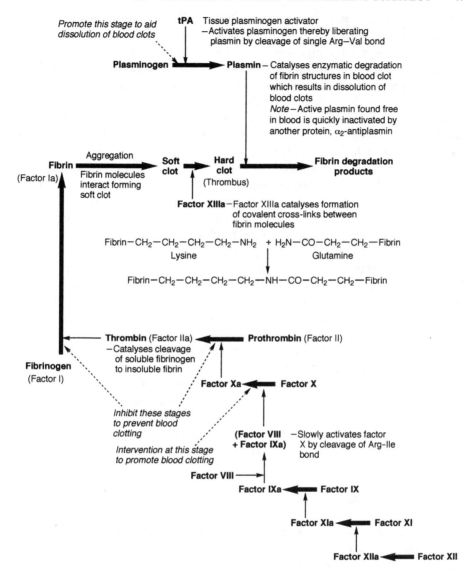

Scheme 2.19 Simplified overview of the fibrin and fibrinolytic cascades. Thick arrows indicate conversion of one component of the cascade into the next, thin arrows indicate the action of the protease or cofactor and dotted arrows indicate stages of the cascade where intervention can be therapeutically beneficial

to haemophilia B (Christmas disease). Such conditions can be effectively managed by replacement therapy of the missing factor. Factors are normally produced by extraction from blood plasma but this source was responsible for the contamination of haemophiliacs with HIV virus present in the blood of certain donors. Partly for this reason, and also because the plasma concentration of the protein is extremely low (200 ng/ml), the recombinant route is becoming the preferred source of factors VIII and IX for clinical use.

Factor VIII is a large glycoprotein – it is composed of 2332 amino acids, with 25 potential glycosylation sites and a molecular mass exceeding 330 kDa. Glycosylation is imperative for this factor to exert its action and mammalian cell expression is the only system which will produce an active product. The factor VIII gene consists of 26 exons or coding sequences distributed over nearly 200 000 base pairs of DNA and its mRNA is approximately 9000 base pairs in length. A full-length cDNA cannot be prepared in the laboratory and so a combination of genomic expression and primed cDNA synthesis is required to obtain the full-length gene. Expression is accomplished by transfecting mammalian systems with plasmids that incorporate the protein-coding sequence under the control of viral promoters and other regulatory signals. Active product is obtained from cell culture.

Factor IX is a complex serine protease with a mass of 56 kDa. It has extensive post-translational modifications, including glycosylation, vitamin K-dependent γ-carboxylation of of the first 12 glutamic acid residues and β-hydroxylation of a single aspartic acid residue. The γ-carboxyglutamic acid residues are essential for the activation of Factor IX and production of this protein in culture again requires mammalian cell expression vectors. The factor IX gene has been cloned from cDNA and expressed in mammalian systems to produce active (carboxylated) Factor IX.

Preparing glycoproteins in mammalian cells (usually Chinese hamster ovary (CHO) cells and baby hamster kidney cells) leads to slight differences when the products are compared to the factors purified from human plasma and some haemophiliac patients can develop an inhibitory antibody response to the recombinant blood-clotting factors.

Anticoagulants and thrombolytic agents

Bleeding disorders such as the haemophilias are relatively rare compared with the reverse situation of hypercoagulability where undesirable clots are formed. Formation of a clot in blood vessels which totally obstructs circulation causes or contributes to conditions such as myocardial infarction (heart attack), pulmonary embolism, deep-vein thrombosis and cerebral stroke. Thrombosis is also a major complication where blood is in contact with foreign surfaces such as prosthetic heart valves. Treatment of thrombotic disorders can be approached either through inhibiting the coagulation cascade using antico-agulants or by potentiating the fibrinolytic system with thrombolytic agents.

Anticoagulants

Most anticoagulants exert their action by inhibiting thrombin. Warfarin (2.11), a coumarin derivative, and the polysaccharide heparin (2.12) are the

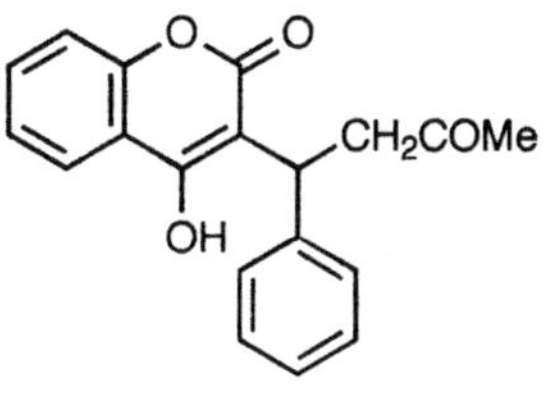

(2.11) Warfarin

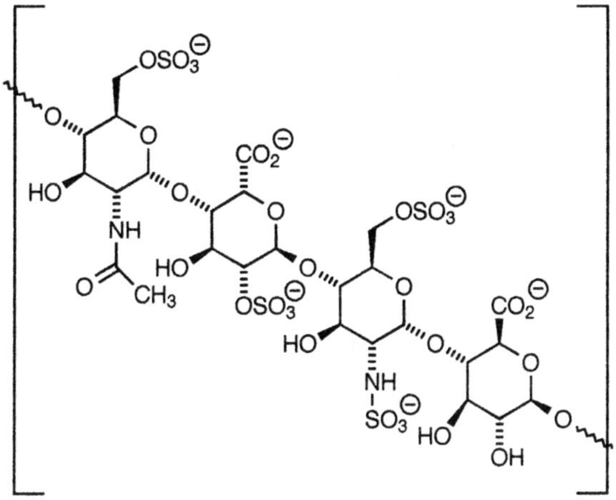

(2.12) Heparin

most widely used anticoagulants. Warfarin interrupts the fibrin cascade by inhibiting post-ribosomal modification of pro-thrombin and several other coagulation factors. Heparin acts indirectly on thrombin by first combining with antithrombin III and then this complex neutralizes several activated factors (the mechanism of action is discussed more extensively in Section 6.3 below). Warfarin and heparin are generally used in a complementary manner; heparin being administered initially or in emergency situations, with warfarin employed for longer-term therapy.

A promising new anticoagulant is the 65 residue, 7 kDa, polypeptide hirudin (2.13). This compound is secreted from the salivary glands of the medicinal leech *Hirudo medicinalis* and is a potent thrombin inhibitor. There are two major isoforms of hirudin which differ in their N-terminal sequences. In hirudin variant 1, from the body of the leech, the sequence is Val–Val, whereas in hirudin variant 2, obtained from the head, it is Thr–Ile. There are also other minor variations. Hirudin is basically a bivalent thrombin inhibitor; the globular N-terminal domain, which is stabilized by three disulphide bonds, binds at the active site region while the extended conformation of the acidic tail-like C-terminus forms numerous salt bridges with the highly electroposi-

$$\begin{array}{l}
\hspace{8.5cm} 10 \hspace{7.5cm} 20 \\
\text{Val-Val-Tyr-Thr-Asp-Cys-Thr-Glu-Ser-Gly-Gln-Asn-Leu-Cys-Leu-Cys-Glu-Gly-Ser-Asn-} \\[4pt]
\hspace{4.5cm} 30 \hspace{7.5cm} 40 \\
\text{Val-Cys-Gly-Gln-Gly-Asn-Lys-Cys-Ile-Leu-Gly-Ser-Asp-Gly-Glu-Lys-Asn-Gln-Cys-Val-} \\[4pt]
\hspace{6.5cm} 50 \hspace{7.5cm} 60 \\
\text{Thr-Gly-Glu-Gly-Thr-Pro-Lys-Pro-Gln-Ser-His-Asn-Asp-Gly-Asp-Phe-Glu-Glu-Ile-Pro-}
\end{array}$$

$$
\begin{array}{l}
\hspace{2cm} 65 \\
\text{Glu-Glu-Tyr-Leu-Gln} \\
\hspace{2cm} | \\
\hspace{2cm} SO_3H
\end{array}
$$

(**2.13**) Amino acid sequence of hirudin isolated from the whole body of leeches. The three disulphide bonds are $Cys^6 - Cys^{14}$, $Cys^{16} - Cys^{28}$ and $Cys^{22} - Cys^{39}$

tive fibrinogen binding site. Purification of large quantities of hirudin from leeches for possible future clinical use is highly impractical but the gene has been cloned from the cDNA and expressed in *E. coli* and yeast to yield a biologically active product.

In recent years, there has been a resurgence in the medicinal use of leeches, particularly in procedures associated with reattachment of fingers and thumbs, where the anticoagulant action of the salival secretion prevents clotting in the very fine blood vessels.

Thrombolytic agents

Anticoagulants function to prevent blood from clotting and are often administered to patients who have suffered heart attacks or strokes to prevent recurrent episodes. Once the three-dimensional fibrin network in a clot has formed, only enzymatic treatment can degrade it. Plasmin, a serine protease, is the principal enzyme involved in the degradation of fibrin. Plasmin itself is quickly inactivated by the plasma protein α_2-antiplasmin and consequently it is not clinically useful as a thrombolytic agent. However, plasmin is derived from the precursor glycoprotein plasminogen by the action of various proteases called plasminogen activators and it is these latter compounds which are of value as therapeutic thrombolytic agents.

(1) Tissue plasminogen activator (tPA). This is a multifunctional serum glycoprotein of molecular weight 70 kDa. It not only functions as a serine protease in specifically converting plasminogen to plasmin, but it also binds to fibrin. If plasmin is generated systemically rather than locally, it can induce the breakdown of other serum factors which can cause bleeding problems. As blood clots are composed of, in part, fibrin, tPA is able to activate plasminogen within the matrix of a clot. This glycoprotein is relatively inactive in the absence of fibrin but once binding occurs the fibrinolytic action increases markedly and this 'clot selectivity' makes tPA a very effective thrombolytic agent. Expression of the cDNA copy of the tPA gene in *E. coli* produces a non-glycosylated but nevertheless active product. Although this protein retains its fibrin-mediated plasminogen activity in the absence of attached oligosaccharide chains, glycosylation may be important for optimizing the performance of tPA as a thrombolytic agent and so mammalian expression systems (CHO cells) are generally favoured for tPA production. The tPA produced by mammalian cells is essentially identical to naturally circulating tPA and thus is not believed to elicit an immunological response. This activator was the first product to be produced and marketed by using mammalian cells instead of bacteria. Transgenic techniques (see Section 2.5 above) have also been used to generate tPA.

(2) Urokinase. This is another serine protease, isolated from human urine or from tissue cultures of human kidney cells, which acts in a similar manner to tPA, i.e. liberates plasmin from plasminogen. Unlike tPA, urokinase is not specific for fibrin. Use of urokinase is limited by its high cost relative to streptokinase (see below) and is used most often in patients with a sensitivity to streptokinase.

(3) Streptokinase. This is a 45 kDa bacterial protein produced from cultures of *Streptococcus haemolyticus*. While streptokinase has no intrinsic proteolytic activity, it forms a 1:1 complex with plasminogen which causes a conformational change on the active site of the plasminogen molecule and renders it proteolytically active. Exposure of the catalytic site of plasminogen in the complex with streptokinase is distinct from that which occurs upon formation of plasmin. The active complex then activates other plasminogen molecules, thereby producing plasmin. However, streptokinase has little affinity for fibrin and it acts systemically. Due to its bacterial origin, streptokinase is immunogenic and its administration can result in adverse allergic reactions. Despite these disadvantages, the ready availability, relatively low cost and reasonable efficacy of streptokinase makes this protein the most commonly used thrombolytic agent. For example, administering intravenous streptokinase within 6 h of a heart attack results in lasting benefit up to 12 months following thrombolysis.

Tissue plasminogen activator, urokinase and streptokinase are all currently prescribed for the treatment of myocardial infarction, deep vein thrombosis, pulmonary embolism and stroke. They are generally administered over relatively short-time periods subsequent to thrombus formation and then this treatment is followed by administration of an anticoagulant for a longer period of time.

Further Reading

Textbook and review articles

- S. P. Adams, Genetic Engineering: Applications to Biological Research, in *Comprehensive Medicinal Chemistry*, P. G. Sammes and J. B. Taylor (Eds.), Vol. 1, P. D. Kennewell (Ed.), Pergamon Press, Oxford, UK, 1990, pp. 409–454.
- P. D. Bailey, *An Introduction to Peptide Chemistry*, Wiley, Chichester, UK, 1992.
- D. Blohm, C. Bollschweiler and H. Hillen, Pharmaceutical Proteins, *Angew. Chem., Int. Ed. Engl.*, 1988, **27**, 207–225.

- W. C. Chan and P. D. White (Eds), *Fmoc Solid Phase Peptide Synthesis: A Practical Approach*, Oxford University Press, Oxford, UK, 2000.
- C. M. Jackson and Y. Nemerson, Blood Coagulation, *Annu. Rev. Biochem.*, 1980, **49**, 765–811.
- J. Jones, *The Chemical Synthesis of Peptides*, Clarendon Press, Oxford, UK, 1994.
- S. B. H. Kent, Chemical Synthesis of Peptides and Proteins, *Annu. Rev. Biochem.*, 1988, **57**, 957–989.
- M. Konrad, Immunogenicity of Proteins Administered to Humans for Therapeutic Purposes, *Trends Biotechnol.*, 1989, **7**, 175–179.
- K. Koths, Recombinant Proteins for Medical Use: The Attractions and Challenges, *Curr. Opin. Biotechnol.*, 1995, **6**, 681–687.
- K. Parfitt (Ed.), *Martindale, The Complete Drug Reference*, 32nd Edn, Pharmaceutical Press, London, 1999.
- J. Sambrook, E. F. Fritsch and T. Maniatis, *Molecular Cloning: A Laboratory Manual*, 2nd Edn, Cold Spring Harbor Laboratory Press, New York, 1989.
- M. D. Taylor, Enzyme Cascades: Coagulation, Fibrinolysis and Haemostasis, in *Comprehensive Medicinal Chemistry*, P. G. Sammes and J. B. Taylor (Eds), Vol. 2, P. G. Sammes (Ed.), Pergamon Press, Oxford, UK, 1990, pp. 481–500.
- W. H. Velander, H. Lubon and W. N. Drohan, Transgenic Livestock as Drug Factories, *Sci. Am.*, 1997, **276** (January), 54–58.
- M. C. Venuti, The Role of Recombinant DNA Technology in Medicinal Chemistry and Drug Discovery, in *Burger's Medicinal Chemistry and Drug Discovery*, 5th Edn, Vol. 1, M. E. Wolff, (Ed.), Wiley, New York, 1995, pp. 661–696.
- J. M. Walker and R. Rapley (Eds), *Molecular Biology and Biotechnology*, 4th Edn, The Royal Society of Chemistry, Cambridge, UK, 2000.
- G. Walsh and D. R. Headon, *Protein Biotechnology*, Wiley, Chichester, UK, 1994.
- I. Wilmut, Cloning for Medicine, *Sci. Am.*, 1998, **279** (December), 30–35.
- At the time of writing (May 2001), the best sites for the human genome project are (a) http://www.hgmp.mrc.ac.uk, (b) http://www.nhgri.nih.gov, and (c) http://www.ncbi.nlm.nih.gov.

Research publications

- E. Atherton, M. Pinori and R. C. Sheppard, Peptide Synthesis. Part 6. Protection of the Sulphydryl Group of Cysteine in Solid-phase Synthesis using N^α-Fluorenylmethoxycarbonylamino Acids. Linear Oxytocin Derivatives, *J. Chem. Soc., Perkin Trans. 1*, 1985, 2057–2064.

- M. Bodansky and V. du Vigneaud, A Method of Synthesis of Long Peptide Chains Using a Synthesis of Oxytocin as an Example, *J. Am. Chem. Soc.*, 1959, **81**, 5688–5691.
- D. V. Goeddel, D. G. Kleid, F. Bolivar, H. L. Heyneker, D. G. Yansura, R. Crea, T. Hirose, A. Kraszewski, K. Itakura and A. D. Riggs, Expression in *Escherichia coli* of Chemically Synthesized Genes for Human Insulin, *Proc. Natl. Acad. Sci. USA*, 1979, **76**, 106–110.
- K. Inouye, K. Watanabe, K. Morihara, Y. Tochino, T. Kanaya, J. Emura and S. Sakakibara, Enzyme-Assisted Semisynthesis of Human Insulin, *J. Am. Chem. Soc.*, 1979, **101**, 751–752.
- R. B. Merrifield, Solid Phase Peptide Synthesis. I. The Synthesis of a Tetrapeptide, *J. Am. Chem. Soc.*, 1963, **85**, 2149–2154.
- J. C. Spetzler and M. Meldal, Evaluation of Strategies for 'One-pot' Deprotection, Cleavage and Disulfide Bond Formation in the Preparation of Cystine-containing Peptides, *Lett. Pept. Sci.*, 1996, **3**, 327–332.
- I. Wilmut, A. E. Schnieke, J. McWhir, A. J. Kind and K. H. S. Campbell, Viable Offspring Derived from Fetal and Adult Mammalian Cells, *Nature (London)*, 1997, **385**, 810–813.
- W. I. Wood, D. J. Capon, C. C. Simonsen, D. L. Eaton, J. Gitschier, B. Keyt, P. H. Seeburg, D. H. Smith, P. Hollingshead, K. L. Wion, E. Delwart, E. G. D. Tuddenham, G. A. Vehar and R. M. Lawn, Expression of Active Human Factor VIII from Recombinant DNA Clones, *Nature (London)*, 1984, **312**, 330–337.
- G. Wright, A. Carver, D. Cottom, D. Reeves, A. Scott, P. Simons, I. Wilmut, I. Garner and A. Colman, High Level Expression of Active Human Alpha-1-Antitrypsin in the Milk of Transgenic Sheep, *Bio/Technology*, 1991, **9**, 830–834.

3

Modification of Endogenous Peptides and Proteins

3.1 Overview

While some naturally occurring peptides and proteins are finding application in the treatment of human diseases, others fall short as drug molecules. Failure of endogenous peptides and proteins to reach the clinic is due to properties intrinsic to these types of molecules. Neuropeptides and peptide hormones, for example, are released and made available to their receptors at the time of physiological need. After exerting a short and intense effort, they are rapidly degraded into their constituent amino acids. Pharmaceutical administration of peptide drugs, on the other hand, takes place far away from the site of action and during transport to their receptors the peptides are continuously being exposed to various enzymes capable of causing their inactivation. Therefore, one of the most important considerations which limits the clinical application of native peptides is rapid degradation by proteolytic enzymes (endo- and exopeptidases). Bioactive molecules exert their biological activity through binding to specific receptor molecules. However, because of the inherent flexibility of peptides they may adopt a number of conformations and be recognized by multiple receptor molecules. This property may lead to undesirable side effects. In an effort to counteract these detrimental properties, modifications are made to the peptide structure and the resulting compounds are known as peptide mimetics or peptidomimetics. Bacterial expression systems are not yet able to incorporate amino acid analogues in a useful way and so the implementation and practicality of the ideas in mimetic design are ultimately dependent on organic synthesis. Many mimetics display improved pharmacological and pharmacokinetic properties such as increased bioactivity, selectivity, metabolic stability, absorption and lower toxicity than those of the parent peptide.

Proteins, like peptides, are not generally orally available and furthermore they are often cleared rapidly from the bloodstream. As a consequence, protein drugs are commonly infused intravenously for several hours to achieve their therapeutic effect. The therapeutic value of pharmaceutical proteins may, however, be improved by chemical and/or genetic means. Specific amino acid residues can be altered at the genetic level by site-directed mutagenesis, the rDNA equivalent of medicinal chemistry. Genetic engineering can also be used to combine domains from different proteins to produce chimeric constructs that incorporate multiple desired properties into a single final molecule. Protein engineering using chemical methods can introduce structural changes that are not encoded by DNA into both recombinant and non-recombinant proteins and often have as their goal the production of a semi-synthetic structure. Genetic- and chemical-modification methods are not mutually exclusive and can be combined to optimize protein-engineering strategies.

3.2   Peptide Mimetics

In the design and development of a peptide mimetic, the aim is to produce a compound which lacks the conformational flexibility and labile amide bonds of the parent peptide but which retains a similar pharmacophore, i.e. the part of the molecule responsible for biological and physiological activity. Developing a peptide mimetic has tended to be predominantly empirical – the endogenous peptide sequence is subjected to structural modification in order to enhance the desirable properties and mimimize the undesirable properties (Scheme 3.1). Modifications include substitution, deletion or insertion of residues, replacing amide bonds with isosteric analogues and introducing conformational constraints. Each analogue is tested in an appropriate assay and in this way structure–activity relationships (SARs) are acquired. This process can be quite time-consuming and expensive but in some cases the structural changes can be based on previous experiments or on data on similar compounds. The elucidation of a peptide's conformation through spectroscopic and molecular modelling studies can provide insight about the structural requirements of the binding receptor. Structural information and data from SARs are then used for the design of more effective peptide mimetics. The cycle of structural modification, synthesis and testing continues until the desired objectives of potency, selectivity, stability, safety and delivery characteristics have been met. Compounds devised from drug discovery programmes then progress through clinical trials.

Receptor proteins are becoming available through molecular biological techniques and elucidation of their structures and analysis of receptor–ligand complexes are now increasingly important elements in drug design and development. In recent years, combinatorial chemistry has entered the

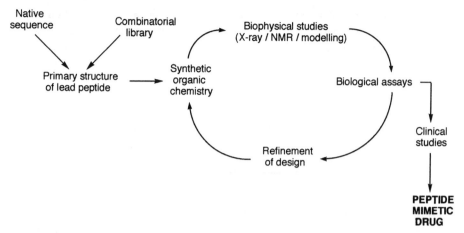

Scheme 3.1 Schematic diagram of the steps involved in obtaining a peptide mimetic drug

drug discovery process. The combinatorial technique involves synthesizing and screening large numbers of compounds in parallel and as such has the potential to be a rapid method of determining whether an initial compound is a lead drug candidate. The limiting factor with this technique is the biological assay and sometimes *in vitro* assays are not confirmed by *in vivo* results.

An important first objective in the development of a peptide mimetic is to delineate which residues are important for activity. This goal is generally accomplished by synthesizing two sets of analogues. The first set of analogues varies from the parent peptide by having shorter chain lengths and this establishes the minimum sequence for complete and partial activities. In the second set of analogues, each amino acid is systematically substituted with alanine along the length of the lead peptide sequence. This process is known as an alanine scan and determines the significance of parameters such as stereochemistry, steric bulk, charge and hydrophobicity at each position. These procedures establish the critical elements of the peptide involved in receptor binding and suggest the positions at which modification might occur to optimize the pharmacological properties. Frequently, only a small number (four to eight) amino acid side-chains are crucial for receptor recognition and the rest of the molecular framework serves to fix the pharmacophore in a particular spatial arrangement.

One notable example where simply truncating the natural sequence resulted in a successful peptide drug is corticotrophin (adrenocorticotrophin (ACTH)) (**3.1**). The latter compound stimulates the synthesis and release of corticosteroid hormones from the adrenal gland. It contains 39 amino acids but it can be reduced to ACTH(1–24)–NH$_2$ (tetracosactide) with only 25% reduction in potency. The first 24 amino acids are responsible for all of the biological activity of ACTH, whereas the remaining residues provide metabolic and structural stability. Synthetic tetracosactide is used clinically to induce an increase in the circulating levels of corticosteroids. Elevated levels of corticos-teroids have proved beneficial in the treatment of inflammatory diseases such as rheumatoid arthritis and asthma. Clinical use of tetracosactide has decreased in recent years as the corticosteroids themselves are now adminis-tered.

$$
\begin{matrix}
& & & & & & & & & 10 & & & & & & & & & & 20 \\
\text{Ser-Tyr-Ser-Met-Glu-His-Phe-Arg-Trp-Gly-Lys-Pro-Val-Gly-Lys-Lys-Arg-Arg-Pro-Val-}
\end{matrix}
$$

Ser-Tyr-Ser-Met-Glu-His-Phe-Arg-Trp-Gly[10]-Lys-Pro-Val-Gly-Lys-Lys-Arg-Arg-Pro-Val[20]-

Lys-Val-Tyr-Pro-Asn-Gly[30]-Ala-Glu-Asp-Glu-Ser-Ala-Glu-Ala-Phe-Pro-Leu-Glu-Phe[39]

(**3.1**) Adrenocorticotrophin (ACTH)

Peptides are flexible molecules which tend to adopt a large number of conformations in solution but assume well-defined conformations when bound to their receptors. The receptor-bound conformation may only be poorly populated in solution but this structure may be promoted by incorpor-ating conformational constraints into the peptide. If the conformation stabi-lized by the constraint closely resembles the structure responsible for the bioactivity, this modification can increase potency and selectivity of the resulting peptide. Some flexibility should be retained in the constrained molecule so that the side-chain pharmacophoric groups may adopt orienta-tions analogous to those in the bioactive conformation of the native peptide. The conformation of a peptide can be stabilized by the introduction of bridges of various lengths between different parts of the molecule. The bridge can either be local and occur within a single amino acid residue or be global and link distant parts of the sequence.

Local constraints

Proline is a naturally occurring constrained amino acid in which the $N-C^{\alpha}$ bond is part of a pyrrolidine ring. Incorporating additional functional groups such as C^{α}-methyl into proline leads to highly constrained mimetics. C^{α} and N alkylation are common steric constraints used to narrow down the conforma-tional space available to a polypeptide chain. α-Methylation severely restricts rotation about the $N-C^{\alpha}$ and $C^{\alpha}-CO$ bonds of the amino acid. Another simple way of modulating the local conformational properties of a peptide backbone is to introduce D-amino acid residues. This is a particularly popular modification because if the analogues are active they would have enhanced stabilities to enzymatic degradation. Structures have also been devised to constrain various types of side-chain functional groups. For example, bridging between N and C^{δ} of Phe and Tyr leads to the constrained amino acids Tic (**3.2**) and HO–Tic (**3.3**) and these are particularly effective for fixing the rotation about the $C^{\alpha}-C^{\beta}$ and $C^{\beta}-C^{\gamma}$ bonds.

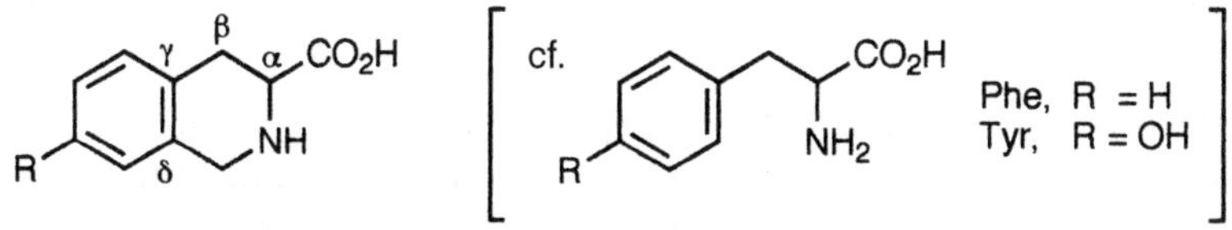

(3.2) 1,2,3,4-tetrahydroisoquinoline-3-carboxylic acid (Tic, R = H)
(3.3) 7-hydroxy-1,2,3,4-tetrahydroisoquinoline-3-carboxylic acid (HO-Tic, R = OH)

Constraints can also be designed to force or stabilize the common architectural elements of peptide structures, i.e. β-turns, α-helices and β-sheets. The majority of the structures mimic β-turns (Figure 3.1) because these are a common motif for many bioactive peptides and the structure is formed from four connected residues (denoted i, $i+1$, $i+2$ and $i+3$) causing a reversal of direction of the peptide chain. A greater number of residues are required to stabilize helices and sheets. Most β-turn mimetics are dipeptide replacements for the $i+1$ and $i+2$ residues at the corners of the turn and structures can either be peptidic or non-peptidic in nature (Figure 3.2). To some extent, turn mimetics have been designed to retain the intramolecular hydrogen-bonding network of an appended peptide chain.

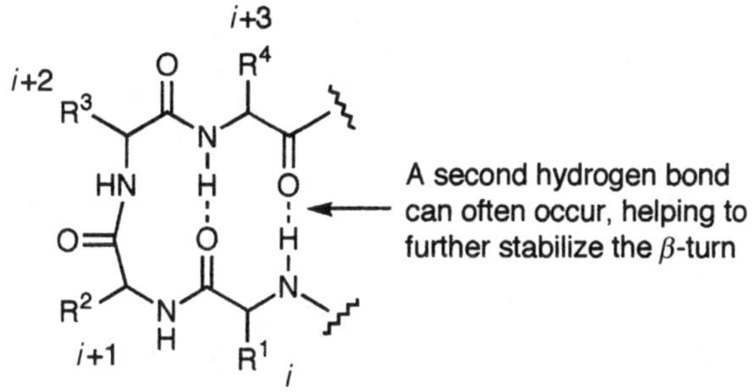

Figure 3.1   Structure of a β-turn

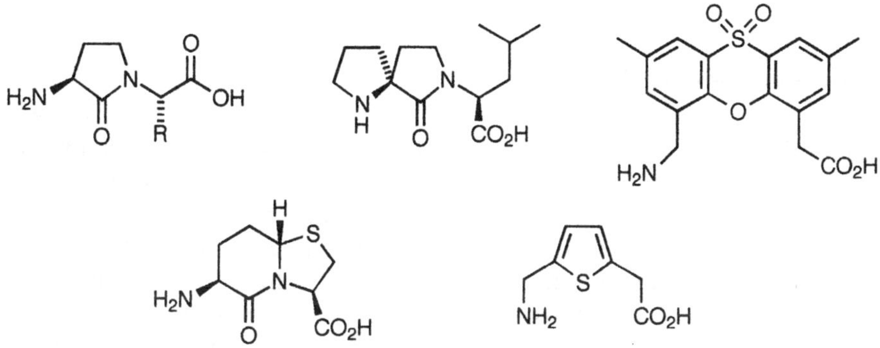

Figure 3.2   Representative β-turn mimetics

Global constraints

Cyclic structures occur in many native peptides such as oxytocin, somatostatin and cyclosporin A. Introduction of a covalent bond linking distant residues of a peptide sequence is most conveniently achieved by forming an amide bond between the N and C termini, between a side-chain and the N or C terminus, or between two side-chains. Alternatively, the constraint can be effected by formation of disulphide bonds between cysteine residues or other thiol-containing amino acids. Cyclization affects the degrees of freedom of all residues within the ring and thus a macrocycle should adopt a more defined conformation than the equivalent linear sequence.

Cyclosporin A (CsA) (**3.4**) is a cyclic undecapeptide produced by the fungus *Tolypocladium inflatum*. It is an immunosuppressant drug (see Section 4.4.2 below) used to help prevent rejection of the donor tissue in patients after transplant surgery. CsA exerts its effect by inactivating the body's T cells, one of the types of white blood cells active in the immune system. This peptide, in addition to being cyclic, has a non-coded amino acid (4-butenyl-4-methyl threonine) and 7 out of the 11 amide bonds are N-methylated, which strongly reduces attack by proteases.

(**3.4**) Cyclosporin A

Amide bond isosteres

Enzymatic resistance is one of the major factors which would prompt the development of a peptide-based drug. Short unmodified linear peptides have serum half-lives in the order of a few minutes and unless they are infused continuously it is necessary to stabilize them against proteolysis. This can be achieved by incorporating unnatural amino acid residues (e.g. D, N-methyl or α-methyl) or amide bond replacements (isosteres) in the positions susceptible to enzymatic hydrolysis. Almost any surrogate of the amide bond (except an ester) will increase resistance towards enzymatic degradation. The most frequently used isosteres for the CO–NH bond are NH–CO (retro-inverso), CH_2–NH (reduced amide), CH_2–S (methylene thioether), CH_2SO (methylene sulphoxide), CH_2–O (methylene ether), CH_2–CH_2 (ethylene, 'carba'), CS–NH (thioamide), (E)-CH=CH (*trans*-olefin), CO–CH_2 (keto methylene) and NH–NR–CO (aza). Most of these modifications are accompanied by changes in geometric structure, electronic distributions and hydrophilic or lipophilic properties. The introduction of amide isosteres also results in local and global changes in dipole moments and in the pattern of intramolecular and peptide–receptor hydrogen-bond formation. Thus, incorporation of amide bond isosteres can not only improve *in vivo* stability as the mimetic is no longer a substrate for peptidases, but can improve selectivity towards receptor sub-types, change pharmacological functions and enhance pharmacokinetic properties.

The retro-inverso modification (NH–CO) is one of the most widely used amide isosteres as, in addition to it being a means of protection against enzymatic cleavage, the amide geometry and topology of the side-chains is largely maintained. Incorporation of amide bond isosteres is not always a successful modification and changes may lead to loss of activity. The carba replacement (CH_2CH_2) is non-polar and does not allow the possibility of intramolecular or peptide–receptor hydrogen bonding, while the reduced amide (CH_2–NH) unit is conformationally different from the amide bond because it does not have any double-bond character. While backbone modification is not always compatible with bioactivity, in some instances this approach had been successful in generating peptides with a prolonged duration of action.

Oxytocin (**3.5**) and vasopressin (**3.6**) are naturally stabilized (via partial cyclization), but can be made more resistant to proteolysis by desamination of the N-terminus or by the incorporation of a D-residue in position 8 such as D-Arg in desmopressin. Demoxytocin (**3.7**) exhibits a higher degree of potency and enjoys a longer circulatory half-life than oxytocin and is sometimes used as an alternative to the native hormone in the induction of childbirth. Vasopressin is an antidiuretic hormone and regulates reabsorption of water by the kidneys. A deficiency in circulating levels of vasopressin results in the

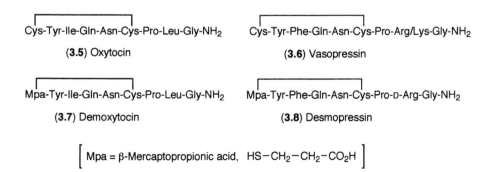

Cys-Tyr-Ile-Gln-Asn-Cys-Pro-Leu-Gly-NH₂

(**3.5**) Oxytocin

Cys-Tyr-Phe-Gln-Asn-Cys-Pro-Arg/Lys-Gly-NH₂

(**3.6**) Vasopressin

Mpa-Tyr-Ile-Gln-Asn-Cys-Pro-Leu-Gly-NH₂

(**3.7**) Demoxytocin

Mpa-Tyr-Phe-Gln-Asn-Cys-Pro-D-Arg-Gly-NH₂

(**3.8**) Desmopressin

Mpa = β-Mercaptopropionic acid,   HS–CH_2–CH_2–CO_2H

onset of a rare form of diabetes known as diabetes insipidus. In the latter, normal renal water reabsorption is impaired and affected individuals experience a constant thirst and excrete large volumes of dilute urine. This condition can be treated successfully by the administration of vasopressin or desmopressin (3.8). Compound 3.8 is often the preferred drug as it has greater antidiuretic activity and has a more prolonged period of action compared with vasopressin.

A peptide mimetic thus embodies the conformational and molecular characteristics thought to be important for biological activity of the native sequence. Mimetics may exhibit enhanced potency and be more selective for various receptor sub-types than their parent sequence but several generations of variants may need to be prepared before a drug candidate emerges.

3.2.1 Mimetics based on luteinizing hormone-releasing hormone

The hypothalamus and pituitary region of the brain can be regarded as hormone headquarters. Luteinizing hormone-releasing hormone (LHRH) (3.9) is one of a number of the hypothalamic peptides which regulate the secretion of pituitary proteins. The releasing hormones are secreted into the hypothalamic–pituitary portal system and travel to the pituitary where they cause the release of pituitary hormones which, in turn, regulate other endocrine glands, for example, the ovaries and testes.

pGlu-His-Trp-Ser-Tyr-Gly-Leu-Arg-Pro-Gly-NH$_2$ [pGlu = Pyroglutamic acid,]

(3.9) Luteinizing hormone-releasing hormone (LHRH)

The decapeptide LHRH causes the release of luteinizing hormone (LH) and follicle stimulating hormone (FSH) from the anterior pituitary and these proteins regulate the reproductive glands. LH stimulates ovulation and secretion of oestrogen in females, while in males it induces testosterone production (Scheme 3.2). Maturation of the follicles in the ovaries and generation of sperm in the testes is stimulated by FSH.

LHRH and the gonadotrophins LH and FSH can be employed therapeutically to treat fertility problems caused by low circulatory levels of the protein hormones. A key feature of LHRH secretion is its pulsatile manner. Mimicking this endogenous secretory pathway by long-term pulsatile delivery (5 to 20 µg over one minute every 90 minutes) of the synthetic decapeptide using an infusion pump can increase the conception rate. Treatment regimens involving direct administration of the gonadotrophins can also be effective in promoting

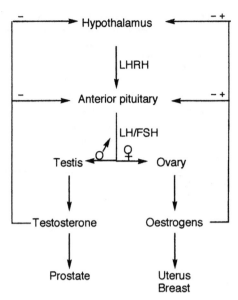

Scheme 3.2   Regulation of the gonads and sex hormone production by the hypothalamus and pituitary glands, where ' + ' and ' − ' represent positive and negative feedback controls, respectively. In men, the feedback is always negative, while in women oestrogens usually modulate gonadotrophin release negatively but a positive feedback occurs prior to ovulation (the 'LH' surge) in each monthly cycle

fertility in anovulatory women. Secretion of hypothalamic and pituitary hormones is regulated by a feedback mechanism – high levels of LHRH desensitizes the LHRH receptor on the pituitary and this decreases LH release in the pituitary. Manipulation of hormonal levels can also therefore lead to contraception –continuous application (daily administration) of LHRH, or more usually one of its analogues which has a longer duration of action, results in cessation of gonadotrophin release. Due to the key role of LHRH in fertility regulation, this decapeptide has been the subject of prodigious research efforts and several thousand analogues have been prepared and assayed for biological activity. Two clinically valuable analogues which have been developed are buserelin (**3.10**) and goserelin (**3.11**).

pGlu-His-Trp-Ser-Tyr-D-Ser(Bu^t)-Leu-Arg-Pro-NHEt

(3.10) Buserelin

pGlu-His-Trp-Ser-Tyr-D-Ser(Bu^t)-Leu-Arg-Pro-AzGly-NH$_2$

(3.11) Goserelin

Table 3.1   Key steps in the development of buserelin and goserelin

Modification/analogue	Potency[a]
• Replacing C-terminal glycinamide (NHCH$_2$CONH$_2$) of LHRH by alkyl amide (NHR) led to (Pro^{9}NHEt)-LHRH	• 3-fold as potent as LHRH in releasing LH and FSH; 6- to 7-fold more potent in inducing ovulation
• Replacing Gly^6 by L-amino acids	• Less potent (0.1–4%)
• Replacing Gly^6 by D-amino acids	• 2–100-fold more potent
• Combination of D-amino acids in position 6 and ethylamide at C-terminus	• More potent than LHRH but effects of two changes are not always additive
• Side-chain-protected amino acids (e.g., D-Lys(Boc), D-Glu(OBu^t), D-Asp(OBu^t), D-Ser(Bu^t)) in (Pro^{9}NHEt)–LHRH	• 40–140-fold more potent in inducing ovulation; 7–20-fold more potent than parent compound (Pro^{9}NHEt)–LHRH
• ⇒ **Buserelin**	
• Replacing Gly^{10} with α-azaamino acid	• Equal in activity in LHRH in inducing ovulation
• Combination of azaglycine residue at position 10 and a D-amino acid residue in position 6, e.g. [D-Ser(Bu^t)^6 AzGly^{10}]–LHRH	• 100-fold more potent
• ⇒ **Goserelin**	

[a]In rats, compared with LHRH.

As can be expected for a linear peptide of its size, LHRH does not have a defined conformation in solution. However, a Tyr^5–Gly–Leu–Arg^{10} type II′ β-turn conformation was suggested from empirical energy calculations. Such turns can be stabilized by the incorporation of D-amino acids at the $i+1$ position, i.e. that corresponding to Gly^6. Several of the peptide bonds in LHRH are susceptible to enzymic degradation *in vivo*, including those of Tyr^5–Gly^6, Gly^6–Leu^7 and Pro^9–Gly^{10}. The two glycine resides in positions 6 and 10 were consequently particularly targeted for modification. Several generations of variants were prepared, with the important compounds leading to the development of goserelin and buserelin being shown in Table 3.1.

Buserelin and goserelin are principally used for the palliative treatment of patients with advanced prostate cancer. Many carcinomas of the prostate are dependent on the male sex hormones (androgens) for their growth and continuous administration of high doses of the superactive agonist of LHRH inhibits androgen-stimulation of prostatic growth through the process of down-regulation of pituitary receptors for LHRH. Hormonal therapy in patients with prostate cancer is never 'curative' but aims to improve the quality of life and prolong survival.

3.2.2   Mimetics based on somatostatin

Somatostatin (**3.12**) is a hypothalamic cyclic tetradecapeptide whose principal action is to inhibit the release of growth hormone from the anterior pituitary.

H-Ala-Gly-Cys-Lys-Asn-Phe-Phe-Trp-Lys-Thr-Phe-Thr-Ser-Cys-OH

(3.12) Somatostatin

The peptide also suppresses the production of the pancreatic hormones insulin and glucagon, has a role in the central nervous system as a neurotransmitter, and is involved in the regulation of gastric secretions. In view of the various actions of somatostatin which have potentially great therapeutic importance, many analogues have been prepared. The peptide itself has a very short duration of action (half-life of less than two minutes) but incorporation of D-amino acids or replacement of the disulphide link by the non-reducible ethylene bridge enhances enzymic resistance and prolongs its activity to approximately three hours. Analogues have also been prepared of somato-statin in an attempt to make its inhibitory effects more specific, and one of these is the octapeptide octreotide (**3.13**).

H-D-Phe-Cys-Phe-D-Trp-Lys-Thr-Cys-Thr-ol

(3.13) Octreotide

Octreotide was designed and developed by Bauer and co-workers at Sandoz but was built on the findings of several other groups. First, not all of the native hormone was necessary for expression of the full activity spectrum and the key pharmacophore was identified as the Phe[7]–Trp–Lys–Thr[10] fragment. Secondly, conformational studies on somatostatin proposed an extended antiparallel β-sheet with the residues Trp[8]–Lys[9] at the corners of a β-turn. Thirdly, replacing Trp[8] by D-Trp gave a compound which was significantly more potent than the native hormone. As theory suggests that a D-amino acid in the $i + 1$ position stabilizes a β-turn, the increased potency of the D-Trp[8] analogue gave further weight to the hypothesis of the existence of a β-turn in the biologically active conformation. Starting with the conformationally constrained mimetic **3.14** of the proposed pharmacophore, the Sandoz group proceeded with systematic modification of the termini. Although the hexapeptide **3.14** had less than 0.001% of the activity of somatostatin in

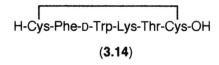

H-Cys-Phe-D-Trp-Lys-Thr-Cys-OH

(3.14)

inhibiting the secretion of growth hormone in rats, addition of a D-Phe residue to the N-terminus gave a heptapeptide which was approximately equal in *in vivo* activity to the native hormone. The aromatic side-chain of this additional amino acid was thought to occupy some of the conformational space available to Phe[6] in somatostatin and also to protect the disulphide bridge from enzymatic attack. Threonine alcohol was identified as the optimum residue at the C-terminus and most likely mimics the conformation of the corresponding Thr[12] in the native hormone. Incorporation of D-Phe and Thr(ol) at the N and C-termini, respectively, culminated in the synthesis of octreotide (**3.13**), a compound which demonstrated the target property in rats of inhibiting the secretion of growth hormone over a therapeutically adequate time-span. Moreover, octreotide was more selective in inhibiting the secretion of growth hormone than that of insulin.

Octreotide has similar properties to somatostatin in humans but a longer duration of action and is used in the treatment of acromegalics and for the management of patients suffering from carcinoid tumours. Such tumours are neoplasms of the diffuse endocrine system and are derived from APUD (amine precursor uptake and decarboxylation) cells. Most of the primary tumours originate in the gastrointestinal tract and produce significant amounts of biologically active peptides, most notably serotonin. Classic symptoms of carcinoid syndrome are flushing, diarrhoea, heart disease and bronchoconstriction. Octreotide does not, however, treat the underlying malignancy of carcinoid syndrome but it is effective in controlling the gastrointestinal manifestations of this disease and patients can enjoy a normal quality of life.

Octreotide can be labelled with a variety of 'radio-metals' which are useful in diagnosis and treatment of cancer. Octreoscan (**3.15**), i.e. octreotide

(3.15) Octreoscan

(3.16)

modified with a chelating group, can form stable complexes with γ-emitting radio-metals such as ^{99m}Tc and ^{111}In and has been used successfully to localize tumours expressing somatostatin receptors. Treatment of tumours rather than diagnosis is possible when particle-emitting isotopes such as the β-emitter ^{90}Y are complexed. Complexes of octreoscan with β-emitters are too labile to be used for therapeutic purposes but the octreotide analogue **3.16** coordinated to ^{90}Y is a stable species and has potential for receptor-mediated radionuclide therapy.

3.3   Chemical Modification of Proteins

Chemical modification of therapeutic macromolecules has a broad application. Modified or conjugated proteins can not only be used directly in the treatment of disease but also indirectly in diagnosis. Antibody–enzyme conjugates are the backbone of enzyme-linked immunosorbent assay (ELISA) procedures, while the attachment of a radio-label on to an antibody molecule is a powerful tool for *in vitro* and *in vivo* diagnosis. The coupling of peptides to carrier proteins is important in antibody production and in vaccine development. *In vitro* laboratory-based applications of chemically modified proteins has had more impact in medical sciences than their *in vivo* use as drug molecules in the clinic.

Modification of proteins requires chemical reactions which are specific for the various types of amino acid side-chain. The most significant amino acids for modification are those containing ionizable side-chains, i.e. aspartic acid, glutamic acid, lysine, arginine, cysteine, histidine and tyrosine. Carboxylate groups of aspartic and glutamic acids and the C-termini can, for example, be esterified while the amino-containing side-chains of lysine, arginine and histidine and the N-termini can be alkylated or acylated. The imidazole ring is also an important reactive species in electrophilic reactions such as in iodination using radioactive ^{125}I (Scheme 3.3). The phenolate ion of tyrosine can also be modified through acylation or alkylation, while its aromatic ring can undergo electrophilic reactions, such as nitration with tetranitromethane, or halogenation with radioisotopes (Scheme 3.3). The most important modifications of cysteine thiol groups in proteins involves formation of reversible disulphide linkages (Scheme 3.4). Native disulphide bonds in proteins (known as cystine residues) can be reduced and the resulting thiols can form disulphide bonds with other thiol-containing compounds, including proteins.

A thiol-exchange reaction is just one method of conjugating two proteins. Coupling can also be effected through the side-chains of lysine, aspartic acid and glutamic acid residues. Amide linkages can be formed by carbodiimide-mediated condensation of the lysine primary amine with the carboxylic acid group of aspartic or glutamic acid. As proteins are generally soluble in aqueous solution, water-soluble carbodiimides such as 1-ethyl-3-(3-dimethyl-

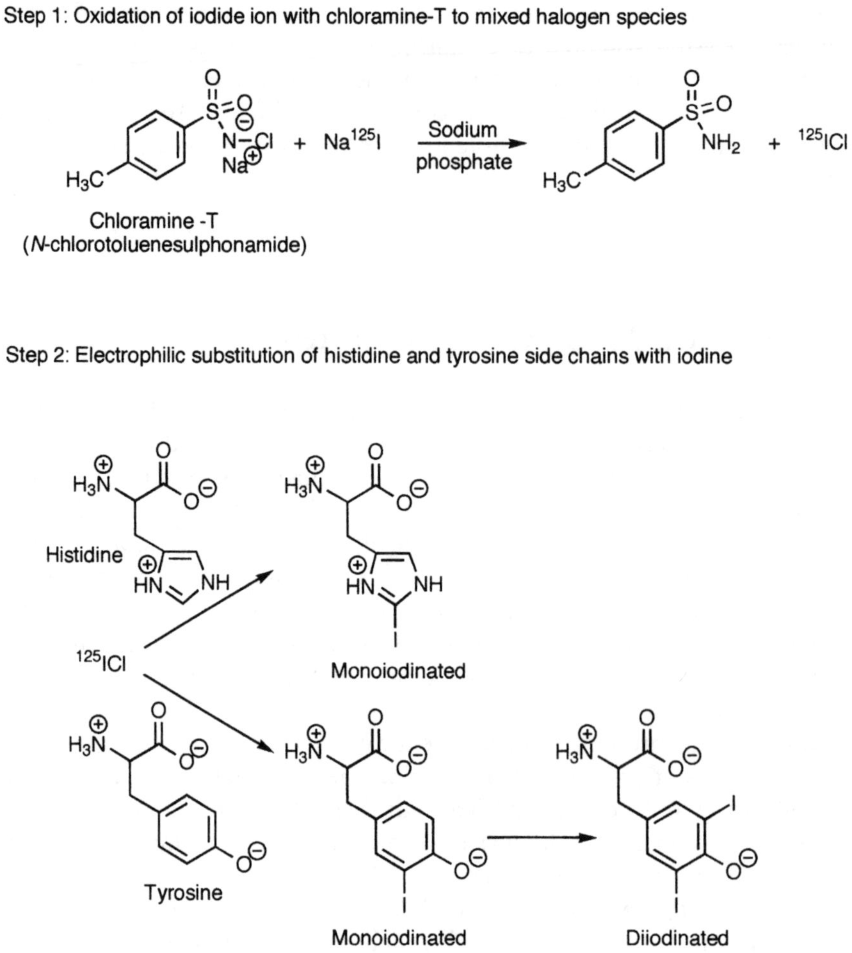
Step 1: Oxidation of iodide ion with chloramine-T to mixed halogen species
H3C
N–Cl
Na
+ Na125I
Sodium
phosphate
NH2
+ 125ICl
Chloramine -T
(N-chlorotoluenesulphonamide)
Step 2: Electrophilic substitution of histidine and tyrosine side chains with iodine
H3N
Histidine
HN
NH
125ICl
Monoiodinated
Tyrosine
Monoiodinated
Diiodinated

Scheme 3.3 Radiolabelling of tyrosine and histidine residues using chloramine-T

aminopropyl carbodiimide hydrochloride (EDC) (**3.17**) are preferred in these bioconjugation procedures in contrast to the DCC used in peptide synthesis (see Section 2.3.1 above) which is soluble only in organic solvents. Amide bond formation in peptide synthesis is carefully controlled by extensive use of protecting groups such that only the desired bond is formed. This is not the case in protein conjugation procedures. Proteins contain many amine and carboxylic acid groups and by carrying out the conjugation reaction in one

$$EtN{=}C{=}N(CH_2)_3NMe_2.HCl$$

(3.17)

$$R-S-S-R' \underset{\text{Oxidation}}{\overset{\text{Reduction}}{\rightleftharpoons}} R-SH + HS-R'$$

Scheme 3.4   Redox reactions between cystine disulphides and cysteine thiols

pot where all of the reagents are present at the same time there is little control over the cross-linking process and many products are formed, only a small percentage of which represents the desired conjugation. Two-step reaction procedures using homo- and heterobifunctional cross-linking reagents (Figure 3.3) can to some extent overcome these shortcomings. These reagents initially react with one protein forming an intermediate which can then be purified before addition of the second molecule to be conjugated. Self-polymerization is still a problem as the first protein has target functional groups on every molecule and these can react with the cross-linker before the second protein is added. Many cross-linking reagents use N-hydroxy succinimide (NHS) esters on one end for coupling to amine groups on the first protein, and maleimide or dithiopyridyl groups on the other end which can react with thiol groups on the second protein (Scheme 3.5). The spacer arm between the two reactive groups at the ends of the cross-linker can be of various length and composition and may be designed to optimize the distance between the two molecules to be conjugated and affect the hydrophobicity of the entire molecule. The structure of the conjugate can be regulated by the degree of the initial modification of the first protein and by adjusting the amount of the second protein added to the final conjugation reaction. Thus, low- or high-molecular-weight conjugates may be obtained to better fashion the product towards its intended use.

3.3.1   Immunoassays

The conjugation of proteins with enzymes is widely used in assay systems. Many assays depend on the ability of antibodies to bind to the antigens that stimulated their initial production. The immune complex can be detected by conjugating either the antigen or antibody to an enzyme, which in turn can convert substrate molecules into easily detectable and quantifiable coloured products. The most important immunoassay, i.e. one employing antibodies, is

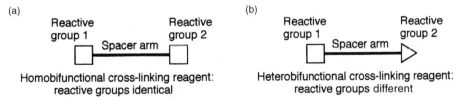

Figure 3.3   Two-step reaction procedures using (a) homo- and (b) heterobifunctional cross-linking reagents. Reproduced from Hermanson, G. T., *Bioconjugate Techniques* 1996, by permission of Academic Press Inc., Orlando, Florida

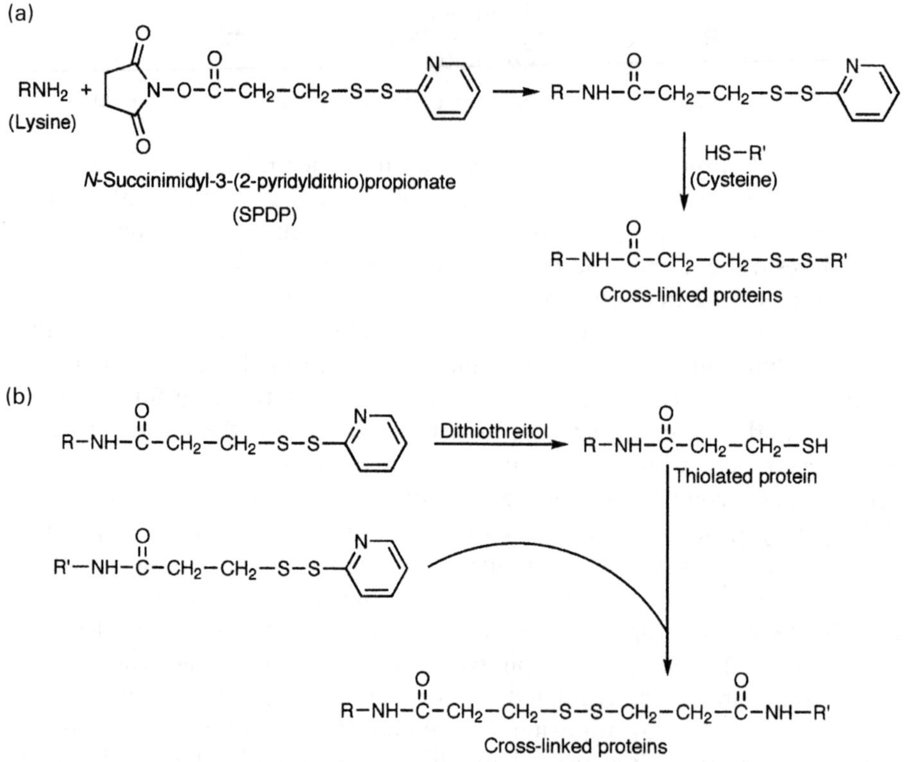

(a)

RNH₂ +
(Lysine)

N-Succinimidyl-3-(2-pyridyldithio)propionate

(SPDP)

R−NH−C−CH₂−CH₂−S−S-

HS−R'
(Cysteine)

R−NH−C−CH₂−CH₂−S−S−R'

Cross-linked proteins

(b)

R−NH−C−CH₂−CH₂−S−S-   Dithiothreitol   R−NH−C−CH₂−CH₂−SH

Thiolated protein

R'−NH−C−CH₂−CH₂−S−S-

R−NH−C−CH₂−CH₂−S−S−CH₂−CH₂−C−NH−R'

Cross-linked proteins

Scheme 3.5   The conjugation procedure for the heterobifunctional cross-linking of an amine-containing protein with a thiol-containing protein. (a) SPDP is one of the most popular heterobifunctional cross-linking reagents and has been used in the preparation of immunotoxins. An antibody reacts with the activated NHS ester end of the SPDP through its amine-containing lysine residues to form a thiol-reactive derivative. The plant toxin ricin contains two subunits (A and B) joined by disulphide linkages. The A chain contains the toxic enzymatic activity and can be isolated from the B chain by reduction with dithiothreitol. The reduced ricin A chain is then mixed with SPDP-activated antibody to effect the final conjugate by disulphide bond formation. (b) SPDP is also effective in creating thiol groups on proteins. Once modified with SPDP, a protein can be treated with dithiothreitol to release the pyridine-2-thione leaving group and form the free thiol group. The thiolated protein is then reacted with a second SPDP-activated protein to create the conjugate

the ELISA system (Figure 3.4). In an ELISA system, antibodies raised against the antigen of interest are adsorbed on to a solid surface, usually the walls of microtitre plate wells. The sample to be assayed is then incubated in the wells. Any antigen present will bind to the immobilized antibodies. After allowing antibody-antigen binding to reach equilibrium, unbound antigen is removed in a washing step. The binding reaction is then detected by the addition of a second antibody which also recognizes the antigen, although at a different site or epitope on the antigen surface from the first antibody. This second antibody

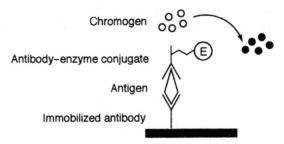

Figure 3.4 Principle of the ELISA technique

is labelled with an enzyme which, after a further washing step to remove any unbound antibody–enzyme complex, can cause a colour change after the addition of the appropriate chromogen.

The enzymes most often used as labels are alkaline phosphatase and horseradish peroxidase. Substrates can yield soluble coloured products, which can be quantified by optical density measurement, or insoluble coloured precipitates. The latter are particularly useful for cell staining. The substrate normally used for alkaline phosphatase liquid-phase assays is p-nitrophenyl-phosphate (PNPP). The latter is hydrolysed by alkaline phosphatase, releasing p-nitrophenol which is yellow and absorbs light at 405 nm. A good substrate that precipitates following the action of alkaline phosphatase is 5-bromo-4-chloro-3-indolyl phosphate/nitro blue tetrazolium (BCIP/NBT). This substrate generates an intense blue–purple precipitate at the site of enzyme binding. The corresponding chromogens normally used with horseradish peroxidase are 3,3′,5,5′-tetramethylbenzidine (TMB), with colour development being measured at 450 nm, and diaminobenzidine (DAB) which gives a brown reaction product.

One of the most common methods of enzyme conjugation uses the homo-bifunctional reagent glutaraldehyde (**3.18**) (Scheme 3.6). The aldehyde residues at both ends of the five-carbon chain react with the ε-NH$_2$ groups of lysines on the proteins forming Schiff-base linkages.

The catalytic properties of enzymes are not the only means of detecting immune complexes in *in vitro* assays. Antibodies can also be labelled radio-isotopes, particularly ^{125}I, through iodination of tyrosine and histidine side-chains (see above) and with fluorescent dyes. One of the most popular of the

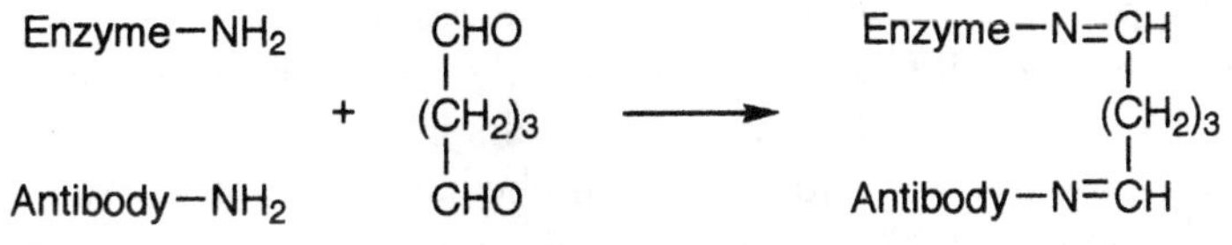

$$\begin{array}{ccc}
\text{Enzyme}-\text{NH}_2 & \text{CHO} & \text{Enzyme}-\text{N}{=}\text{CH} \\
& | & | \\
+ & (\text{CH}_2)_3 & \longrightarrow \quad (\text{CH}_2)_3 \\
& | & | \\
\text{Antibody}-\text{NH}_2 & \text{CHO} & \text{Antibody}-\text{N}{=}\text{CH} \\
& \textbf{(3.18)} &
\end{array}$$

Scheme 3.6 Coupling of enzyme to antibody by using glutaraldehyde

fluorescent dyes is fluorescein isothiocyanate (**3.19**) which reacts under alkaline conditions with primary amines in proteins to form conjugates with an isothiourea linkage.

(3.19) Fluorescein isothiocyanate

Assays based on the use of antibodies have widespread application in the monitoring of extraction and purification protocols of these molecules and in diagnosis. Immunoassays are used for a broad variety of diagnostic applications, including assessment and monitoring of various cancers, detection of specific hormones and pathogen identification and monitoring.

3.3.2   Modification with synthetic polymers

Lysine-specific reagents are widely used for protein modification. Exposed lysine residues are relatively abundant in proteins, the ε-NH$_2$ group is reactive, and acylation and alkylation reactions are rapid, occurring in high yield to give stable amide or secondary amine bonds. Modification of lysine-containing proteins with synthetic polymers such as poly(ethylene glycol)s (PEGs) can be of benefit for *in vivo* applications. Poly(ethylene glycol) (**3.20**) consists of repeating units of ethylene oxide that terminate in hydroxyl groups on either end of a linear polymer chain. The hydroxyl groups can be activated to allow coupling of the polymer to lysine residues in proteins. Since the polymer backbone of PEG is not of biological origin, it is not readily degraded by mammalian enzymes. This property results in the slow degradation of the polymer when used *in vivo* thus extending the half-life of modified substances. PEG modification creates a new surface for any protein to which it is coupled. This may mask antigenic determinants, thus protecting the 'pegylated' molecule from being inactivated by antibodies in the blood stream. The price of an increased plasma half-life and a reduction in immunogenicity conferred by PEG modification is usually a decrease in biological activity.

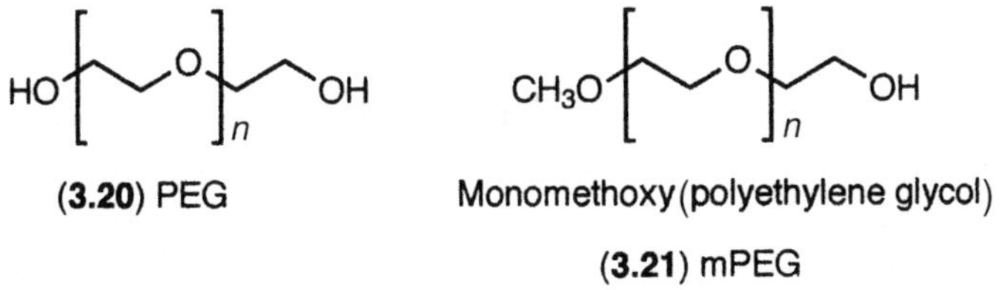

(3.20) PEG

Monomethoxy(polyethylene glycol)

(3.21) mPEG

PEG is a bifunctional polymer and its conjugation to proteins through both hydroxyl groups can cause cross-linking and polymerization of modified molecules. These problems can be avoided by using the monofunctional PEG polymer (mPEG) (**3.21**). One end of the mPEG chain is blocked with a methyl ether group and so activation of the polymer forms a monovalent intermediate that can be coupled to proteins without cross-linking and polymerization. Several approaches can be used to form the activated polymer. One of the most frequently used methods involves reacting mPEG with *N*-hydroxysuccinimidyl chloroformate (**3.22**) to give the succinimidyl carbonate derivative **3.23** (Scheme 3.7). Reaction of this intermediate with lysine-containing proteins generates the 'pegylated' species. The modified protein contains stable carbamate linkages and so will not lose PEG by hydrolytic cleavage.

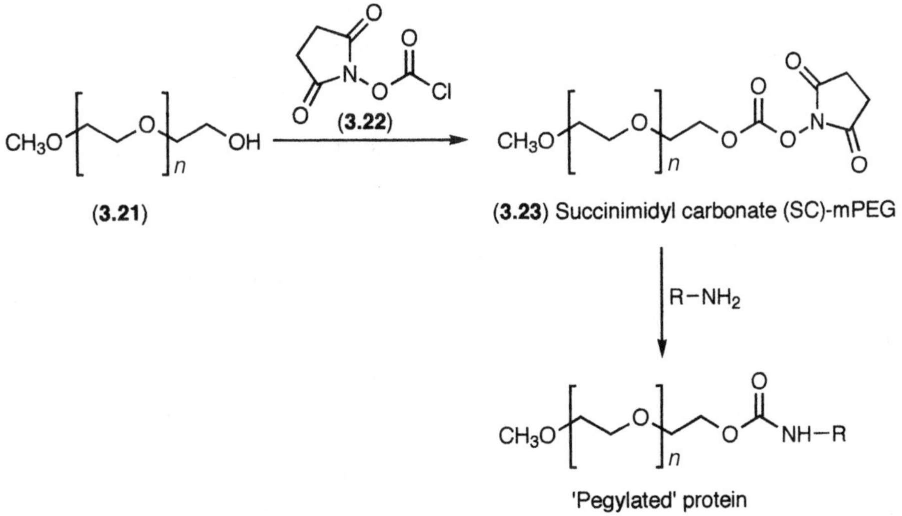

Scheme 3.7 PEG modification of proteins by succinimidyl carbonate activation

Application of the PEG derivatization technique to adenosine deaminase (ADA) improves its pharmacological profile – the half-life in plasma increases and the immunogenicity decreases. One of the approved methods of treating the immunodeficiency condition, resulting from an inherited deficiency of the adenosine deaminase enzyme, is administration of PEG-bovine ADA (see also Section 5.2.3 below).

Interleukin 2 (IL-2) (see Sections 4.4.1 and 4.5.2 below) is a cytokine which stimulates the proliferation of T cells and thus amplifies the immune response to an antigen; it also has actions on B cells and induces the production of interferon-γ and the activation of natural killer cells. IL-2 is used in the adoptive immunotherapy of some cancer patients and is usually given by intravenous infusion of one of its recombinant forms. PEG-IL-2 is more soluble and has a much longer plasma half-life than the native protein and can be administered as an injection rather than by continuous infusion.

The balance between improving pharmacokinetic properties and the reduction in specific activity has not proved favourable in the case of PEG modification of tPA. Attachment of PEG to tPA, a fibrinolytic protein used as in the treatment of acute thrombolytic disorders (see Section 2.6.2 above), resulted in the loss of the enzyme's biological activity and illustrates the unpredictability of the chemical modification procedure.

3.3.3   Modification of enzyme active sites

Many reagents tend to modify all accessible groups of a given type and it can be difficult to alter a specific residue. One situation where it is possible to effect selective modification of functional groups is to target the catalytic centres of enzymes. Amino acids in the active sites of enzymes have unusual reactivities and this makes them distinguishable from other residues of this type. In the serine protease hydrolysis of peptide bonds (Scheme 3.8), for example, a serine hydroxyl group in the active site reacts with the substrate to form an acyl–enzyme intermediate **3.24**. Alcohols are normally poor nucleophiles but a hydrogen-bonding network between aspartic acid, histidine and serine residues in the active site confers enhanced nucleophilicity to the serine hydroxyl group, thus permitting its reaction with the carbonyl group of the scissile-peptide bond and the formation of an acyl–enzyme complex. This acyl–enzyme mechanism can be exploited to synthesize stabilized and isolable acyl–enzyme intermediates.

The enzyme components of the coagulation and fibrinolytic pathways are serine proteases and acyl-group modification of the active site of the $1:1$ plasminogen–streptokinase complex (see Section 2.6.2 above) has proved to be therapeutically beneficial. The active centre of streptokinase can be specifically acylated with a p-anisoyl group ($-CO-C_6H_4-OCH_3$) by reaction with the reactive ester **3.25**. The presence of the p-anisoyl group prolongs the

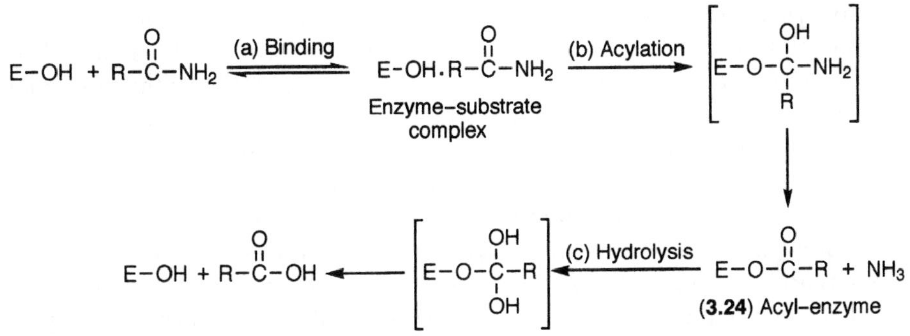

Scheme 3.8 Serine protease-catalysed hydrolysis of amide substrates. (a) The enzyme and substrate first associate to form a non-covalent enzyme–substrate complex held together by physical forces of attraction. (b) Attack of the hydroxyl of serine on the substrate gives the first tetrahedral intermediate. This intermediate then collapses to give the acyl–enzyme **3.24**, releasing the amine. (c) The acyl–enzyme then hydrolyses to form the enzyme-product complex via a second tetrahedral intermediate. Ester substrates can be hydrolysed in a similar manner

plasma half-life of streptokinase and prevents the enzymatic action of the streptokinase–plasminogen complex. The inactive complex can still bind to a blood clot as the fibrin binding sites on plasminogen are separate from the catalytic centre. Once bound, the complex becomes activated by slow deacylation and the plasminogen-activating enzyme is regenerated. Anisoylated lysplasminogen streptokinase activator complex is a valuable treatment for myocardial infarction. The enzyme complex is noted not only for its clinical efficacy – heart function improves and mortality decreases – but also for its ease of administration which promotes early intervention. It can be given as a simple rapid intravenous injection rather than as a prolonged infusion which is required in the case of streptokinase itself.

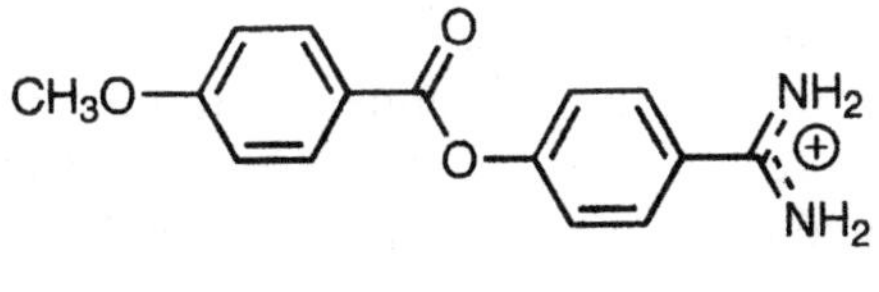

(3.25)

3.4 Protein Engineering

Chemical modification of proteins is restricted to the chemically reactive side-chain functional groups and the reactions, while specific for residues of a certain type, may not be selective for a particular amino acid residue. Modification of residues at a given location in a protein can be achieved by manipulating the corresponding gene. Manipulation of DNA at the molecular level is known as site-directed mutagenesis (see Section 3.4.1 below). Three types of mutation can be made, namely replacements, insertions and deletions of residues. Incorporating a single or double-base mismatch into the nucleic acid sequence changes the codon for the target amino acid into the codon for the desired mutant residue, inserting sequences of DNA adds amino acid residues and deleting sections of the gene removes sections of the polypeptide chain. The basic method of genetic engineering is flexible and can be adapted for more complex mutagenesis applications e.g. the mutagenesis of the six amino-acid sequences of an antibody's heavy- and light-chain variable regions during antibody humanization (see Section 3.4.3 below). Addition or removal of protein domains is another productive route to protein engineering (see Section 3.4.2 below).

3.4.1   Site-directed mutagenesis

A number of methods are available to alter a defined nucleotide sequence within a gene in a precise manner. In the original method, a synthetic oligonucleotide incorporating the required mutation is hybridized to a circular single-stranded DNA template. A complementary strand is then synthesized by DNA polymerase and deoxynucleotide triphosphates (dNTPs) using the oligonucleotide as a primer. DNA ligase is then used to seal the new strand to the 5′-end of the oligonucleotide (Scheme 3.9). The double-stranded DNA, homologous except for the intended mutation, is then amplified *in vivo* to give mutant and wild-type progeny. The clones containing mutant DNA can be identified by any of a number of screening and selection procedures. This oligonucleotide-directed mutagenesis procedure generally results in low mutational efficiency and a large number of clones have to be screened in order to obtain one with the required sequence.

Current practice for the creation of defined mutations now tends to focus on the application of two techniques, namely cassette and PCR-based mutagenesis. In the first, two complementary strands of DNA containing the desired mutation are synthesized and annealed to generate a duplex cassette. The duplex DNA is then inserted between two restriction endonuclease recognition sites on a piece of plasmid DNA (Scheme 3.10). The mutagenic plasmid is then transformed in *E. coli* in the normal manner. In the second approach, the desired modification is amplified *in vitro* by means of the PCR reaction. The latter is a method which enables amplification of DNA (see Scheme 2.10 above). The method involves repeating a series of three steps: (1) denaturation of the template DNA strands, (2) annealing of primers and template to form

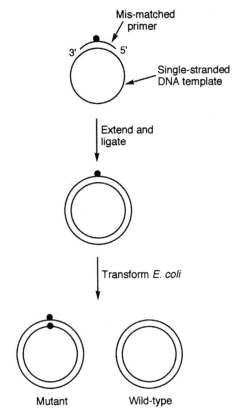

Scheme 3.9 Oligonucleotide-directed mutagenesis

duplex DNA, and (3) extension of the primers from their 3'-ends by a DNA polymerase. In PCR mutagenesis, one of the oligonucleotide primers employed in the amplification procedure has one or more sequence mismatches to the template so that on amplification a mutation is fixed in the progeny DNA (Scheme 3.11). The amplified DNA is then linked to a bacterial-replication origin site in a plasmid and cloned by introduction into bacteria.

Expression of the mutant gene leads to the production of the protein variant which is then assessed to evaluate the effect of the alteration. Substitution of the wild-type amino acids by residues with varying hydrophobicity, chemical-bonding potential and size can affect the potency, efficacy, bioavailability and/or metabolism of the protein. Modifications to proteins at the genetic level are limited to the repertoire of the 20 natural amino acids encoded by DNA, as expression systems are not yet able to incorporate amino acid analogues in a useful way.

2 complementary
strands of
DNA containing
desired mutation
are synthsized
and annealed to
generate duplex
cassette

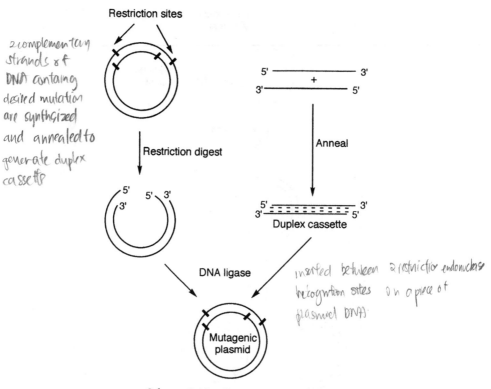

inserted between 2 restriction endonucleas
recognition sites on a piece of
plasmid DNA.

Scheme 3.10   Cassette mutagenesis

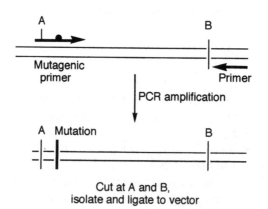

Cut at A and B,
isolate and ligate to vector

Scheme 3.11   PCR mutagenesis

3.4.2   Chimeric toxins

Chimeric toxins are composed of a protein toxin linked to a binding ligand
such as a growth factor or antibody. These molecules are being developed as a
treatment option for cancer. The basic concept is simple; the ligand of the

chimeric toxin binds to the cell surface, is internalized and then the toxin kills the cell directly by inhibiting protein synthesis or indirectly by inducing apoptosis. First-generation chimeric toxins were composed of whole antibodies chemically conjugated to plant toxins, such as ricin, or bacterial toxins, such as the *Pseudomonas* exotoxin (PE) and diphtheria toxin (DT) (Figure 3.5(a)). However, the main problems with these conjugates was their non-specific toxicity towards non-cancer cells and their large size which resulted in poor tumour penetration. Using protein engineering techniques, molecules have been designed and constructed with much higher specificity for cancer cells and with greatly reduced toxicity towards normal cells.

The toxins by themselves bind and kill normal cells. A common feature of these potent toxins is their catalytic activity, i.e. only a few molecules need to reach the interior of the cell to achieve devastating results. Most toxins have a modular make-up; one domain binds to the receptor on the outside of the cell, another assists in traversing the plasma membrane, while a third region contains the catalytic activity (Figure 3.5(b)). Removing the binding domain and replacing it with a ligand specific for cancer cells should produce cytotoxicity in only those cells to which the ligand can bind. Cancers are becoming more definable by proteins displayed on the malignant cell surface, and then, by directing the toxin selectively to these surface proteins it is, in principle, possible to selectively kill cancerous cells while sparing normal cells. The specificity of the antibody–antigen interaction makes immunoglobulin

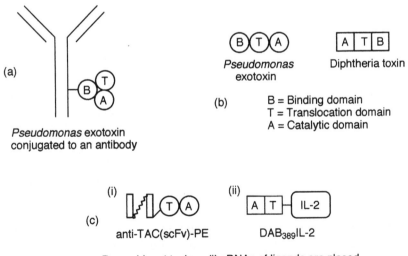

(a)

Pseudomonas exotoxin conjugated to an antibody

Pseudomonas exotoxin

Diphtheria toxin

(b)

B = Binding domain
T = Translocation domain
A = Catalytic domain

(c)

(i) anti-TAC(scFv)-PE

(ii) DAB$_{389}$IL-2

Recombinant toxins. (i) cDNAs of ligands are placed at the 5' end of the truncated PE gene. (ii) Binding by DT is mediated by sequences at the 3' of its structural gene and therefore ligands are added to truncated form of DT at the 3' end

Figure 3.5   Chimeric toxins

molecules ideal targeting agents. Antibodies are large (approximately 150 kDa) proteins, but the antigen binding specificity of an antibody is mediated by only a small part of the total molecule, namely the variable regions (Fv). The Fv fragments are not very stable but their two component units (V_H and V_L) can be connected by a small peptide to produce a stabilized single-chain antibody, the scFv (see Section 3.4.4 below). Moreover, reducing the size of an antibody molecule increases its tumour penetration potential.

Recombinant chimeric toxins are constructed by fusing modified toxin genes to DNA elements encoding growth factors or scFv antibodies. Such fusion genes are then placed in plasmids together with control elements for production of the recombinant protein (Figure 3.5(c)) in *E. coli*. Chimeric toxins can also be produced by chemical coupling methods but large amounts of protein and toxin are required and furthermore the conjugation methods often produce heterogeneous products.

One of the most potent of the antibody-derived chimeric toxins against leukaemia is anti-TAC(scFv)-PE (Figure 3.5(c(i))) which targets the IL-2 receptor α-sub-unit (TAC antigen) which is overexpressed on many (T-cell) leukaemias. The pre-clinical development of this chimeric toxin has been completed and clinical trials have been initiated to evaluate its use in patients with IL-2-receptor-expressing leukaemias, lymphomas and Hodgkin's disease. Denileukin difitox (DAB_{389}IL-2, Figure 3.5(c(ii))) containing the initiator methionine, the first 388 amino acids of DT and human IL-2 also produces specific cytotoxicity in cells expressing the IL-2R and this chimeric toxin has recently being licensed for use in the treatment of patients with cutaneous T-cell lymphoma.

3.4.3   Chimeric and humanized antibodies

Monoclonal antibodies (mAbs) can be used for the treatment of disease. Monoclonal antibodies have specificity for only one antigenic epitope and are prepared by using hybridoma technology (see Section 4.2.2 below for further discussion). This procedure is essentially simple, namely B lymphocytes from the spleen of an immunized rodent are mixed with a continuously proliferating B-lymphoma cell line, and their membranes fuse to give a single antibody-secreting cell (hybridoma) that can proliferate indefinitely. The clinical utility of mAbs is, however, limited because they are generated in mouse cells. Rodent mAbs have a short survival time in humans and also have the potential to generate an immune response (human anti-mouse antibodies or HAMAs) in humans which neutralizes their therapeutic effect. Furthermore, the responses induced by murine antibodies are poor because they only weakly recruit human effector functions. Attempts to use hybridoma technology for generating human mAbs have been hampered by the lack of a suitable immortal cell line. However by cloning the cDNAs encoding the mAb it is possible to alter the sequence of rodent antibodies and make them less immunogenic in humans.

The engineering of antibodies has been facilitated by the modular arrangement of the protein domains. A typical antibody molecule is composed of two heavy and two light chains and both chains have distinct constant and variable regions (Figure 3.6). The variable domains at the amino-terminal ends of the heavy and light chains form the antigen-binding sites while the constant domains of the heavy chains (mainly C_{H2} and C_{H3} determine the other biological properties of the molecule. Each antibody domain is encoded by a different genetic exon and to build recombinant antibodies these exons are added together.

There are two types of genetically engineered antibody variant: (1) chimeric antibodies with mouse variable regions and human constant regions, and (2) humanized antibodies which are in essence the antigen-binding complementarity-determining regions (CDRs) of the parent rodent monoclonal antibody in association with human framework regions.

(1) Chimeric antibodies

human antimouse · Ab

Most of the HAMA response to mouse antibodies in humans is directed against the constant portion of the mouse antibodies mAbs. By replacing these regions by their human counterparts, the mAb can be made to appear more human and be less immunogenic in patients. Chimeric antibodies are created by transplanting the variable domains of a rodent antibody to the constant

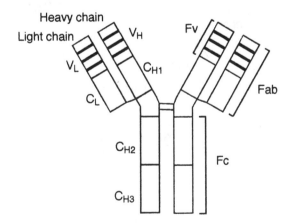

Figure 3.6 Schematic representation of an antibody molecule: C_L, C_{H1} to C_{H3}, constant regions; V_L, variable region of the light chain; V_H, variable region of the heavy chain; Fab, fragment antigen binding comprising the variable Fv region ($V_L + V_H$) and part of the constant region ($C_L + C_{H1}$); Fc, fragment constant ($C_{H2} + C_{H3}$) × 2; CDRs, complementarity-determining regions (shown in bold)

domains of human antibodies. The gene segments encoding the variable domains (V genes) are isolated from the mRNA of a culture of hybridoma cells and amplified by using the polymerase chain reaction. Appropriate restriction sites are introduced at each end of the variable regions' cDNA and then the V genes are spliced into expression vectors which contain genes encoding the constant domains (Scheme 3.12). Separate heavy and light chain transfection vectors are constructed and they also include effective promoter and enhancer elements and antibiotic selection markers. The vectors are normally expressed in mammalian hosts as both complement and cell-mediated killing require fully glycosylated antibody.

Chimeric antibodies have therapeutic advantages over mouse antibodies – they have extended serum half-lives in humans and, with approximately

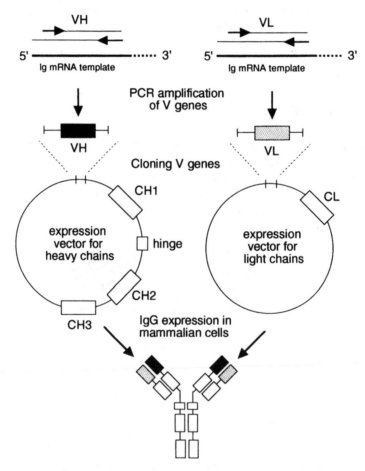

Scheme 3.12 Building chimeric antibodies: (a) PCR amplification of V genes; (b) cloning V genes; (c) antibody expression in mammalian cells (bold arrows represent primer oligonucleotides). Reprinted from *TIPS*, **14**, G. Winter and W. J. Harris, Antibody-based Therapy, 139–143, copyright (1993), with permission from Elsevier Science.

60–70% of the molecule being human, the HAMA response is reduced, although not eliminated. As the constant region now stems from man, chimeric antibodies activate some helper functions, such as the antibody-dependent cellular cytotoxicity, of the immune system more efficiently than the mouse antibodies.

(2)   Humanized antibodies

In humanized antibodies, the antigen-binding loops of a rodent antibody have to be recreated within a human framework. Each antibody V-domain consists of a β-sheet framework structure surmounted by the antigen-binding loops (the CDRs) (Figure 3.7). By transplanting the CDRs from rodent mAbs to human antibodies, the antigen-binding site is also transferred. While the antigen-binding site is formed primarily by the CDRs, some framework residues also contribute to the CDR conformation and antigen binding. These specific amino acids must be identified and retained in the humanized antibody. Furthermore, the mode of association of the light- and heavy-chain variable domains often changes upon antigen binding. Computer generated models and/or an X-ray structure of the Fv region are used to predict the additional changes that need to be introduced into the human framework to produce a recombinant antibody with the required antigen-binding affinity.

The initial steps in humanizing a mouse antibody are similar to those used in the creation of a chimeric antibody. Hybridoma RNA is isolated and cDNA is synthesized by using V region primers, dNTPs and reverse transcriptase

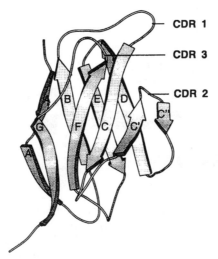

Figure 3.7   Variable domain structure showing the β-sheet framework and the three CDR loops of one antibody chain. Reproduced from Rees, A. R. *et al*, "Antibody Design: Beyond the Natural Limits", *Trends in Biotechnology*, **12**, 199–206, 1994 with permission of Elsevier Science.

(Scheme 3.13). The V region genes are then amplified by PCR and, by using suitable primers, restriction sites are incorporated to allow for subsequent cloning into vectors. In humanization protocols, the cloned DNA is then sequenced and the nucleotide sequence is translated into the corresponding amino acid sequence. Comparison with the database of antibody amino acid

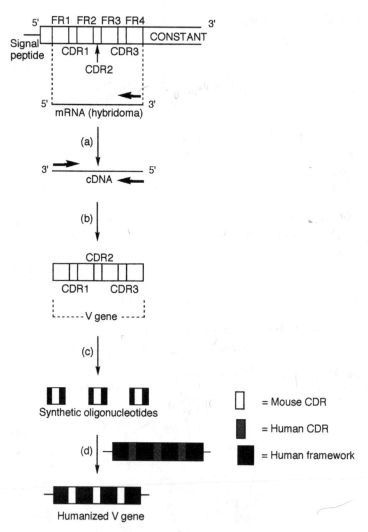

Scheme 3.13 Strategy for humanizing antibodies. (a) RNA extracted from hybridoma cells is primed with an oligonucleotide (represented by a bold arrow) specific for the 3′ end of the variable region and cDNA is synthesized. (b) Amplification of the cDNA of the heavy and light chains using PCR. (c) The amplified DNA fragments are cloned into vectors and the V genes are then sequenced. Oligonucleotides encoding the mouse CDRs, and extended by nucleotides (represented by black bars) to promote efficient hybridization to the human single-stranded template, are synthesized. (d) The mouse CDRs are placed into the human framework by oligonucleotide-directed mutagenesis to produce the humanized V genes

sequences allows the CDR regions to be delineated and computer graphic analysis of the protein structure is used to identify critical residues from the murine sequence for inclusion within the human framework. Six oligonucleotides are synthesized which span the CDR coding region and include any additional exchanges in the adjacent framework regions. At their 5′ and 3′ regions they contain about 15 perfectly matched nucleotides which are complementary to the flanking framework regions of the human V genes. These sequences are then grafted into the human V_L and V_H framework by using oligonucleotide-directed mutagenesis (see Section 3.4.1 above). Mutagenesis of the three CDRs in a human V region DNA sequence can be achieved by three sequential single reactions or in one round of mutagenesis by simultaneously annealing all three mutagenic oligonucleotides. The humanized V_H and V_L genes are then cloned into the expression vectors and the vectors cotransfected into mammalian cells as described in (1) above.

Campath-1, a murine antibody directed against the CDw52 antigen of human lymphocytes, was the first antibody to be humanized. Grafting just the CDRs failed to transplant the binding affinity to a human antibody but by transferring a few murine framework residues, which were identified by molecular modelling to be important for antigen binding, the binding affinity was restored. The humanized antibody Campath-1H has been used for the treatment of non-Hodgkin's lymphoma and rheumatoid arthritis. Evidence of clinical benefit was observed but with repeated therapy over half of the rheumatoid arthritis patients developed HAMA responses and the use of Campath-1H as a treatment for rheumatoid arthritis has now been halted. Trastuzumab is a humanized anti-p185^{HER2} antibody which has been shown to be effective in limiting the growth of tumours in breast cancer patients. Humanization of the murine mAb directed against the cell surface protein p185^{HER2} again involved transferring not only the antigen-binding loops from the murine antibody into human V domains but also several framework region residues as directed by molecular modelling.

Chimeric and humanized antibodies are currently used in the treatment of a range of conditions including cancer, transplant rejection and a variety of immunological conditions such as rheumatoid arthritis and Crohn's disease (see also Section 4.2.2 below). Many more antibodies, particularly the humanized forms, are in clinical development.

3.4.4   The single-chain Fv

In cancer immunotherapy the fraction of injected mAb that effectively reaches the tumour is usually less than 1% of the injected dose. The poor penetration of whole antibodies (mass of approximately 150 kDa) into solid tumour mass is one of the greatest barriers to their use in effective cancer therapy. One method of improving this is to use small antibody fragments. A popular format is the single-chain Fv (scFv) consisting of the V_H and V_L domains tethered by a flexible linker such as the 15 amino acid peptide $(Gly_4-Ser)_3$ (Scheme 3.14(a)). Single-chain Fv proteins consist of about 250 amino acids and have molecular weights of approximately 26 to 27 kDa. The low molecular mass of scFvs causes them to be cleared much faster from the circulation than Fab fragments ($t_{1/2}$ in blood approximately 2 h compared with 14 h for Fab) and this is likely to make them less immunogenic than larger fragments or whole antibodies. Constructing a scFv involves connecting the cDNAs for the V_L and V_H domains (isolated from the mRNA of a culture of hybridoma cells and amplified using PCR) by an oligonucleotide sequence that codes for the appropriate linker and then expressing the protein in bacterial hosts. Sometimes, the antigen binding capability of the scFv is altered compared with the Fab fragment or whole antibody and it could be that, in common with humanized antibodies, constant region residues are important in maintaining the correct CDR conformation. Nevertheless, single-chain Fvs have considerable promise in medicine. While scFvs cannot recruit effector cells to produce cell killing, they can be made as genetic fusions to produce fragments bearing a variety of cytotoxic molecules such as toxins (Scheme 3.14(b)). The scFv proteins may also be used for imaging or therapy by labelling with radioisotopes.

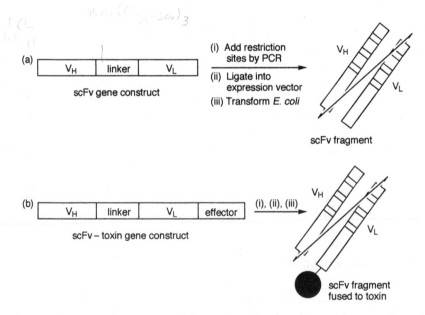

Scheme 3.14 Expression of scFv antibody fragments in bacteria. (a) Single chain Fv polypeptide chain. There are two possible permutations, i.e. V_L-V_H and V_H-V_L (shown). (b) scFv fusion protein, whereas the effector domain can be fused to either the amino or carboxyl (shown) termini of the scFv polypeptide chain

3.4.5   *Phage display*

While humanization can be highly successful, the resulting antibody still contains approximately 50 to 60 residues of mouse antibody sequence which can be recognized as foreign and raise a human immune response. Humanization is also a laborious procedure requiring alternating rounds of computer modelling and genetic modification to retain the specificity and affinity of the original mouse antibody sequence. It would be preferable to produce fully human antibodies as human therapeutics. Making human antibodies by the hybridoma route is fraught with difficulties. The use of mouse myeloma as a fusion partner for human cells leads to preferential loss of human chromosomes and instability of the hybrids. The immunization of humans with toxic antigens is neither practicable nor ethical and, even when an immunized donor can be found, the production of mAbs is still technically demanding. The primary source of cells for generating antibodies from humans are peripheral blood lymphocytes but these contains few cells actively involved in the immune response.

Correctly folded antibody fragments, e.g. Fab, Fv or scFv, can be expressed in *E. coli* by routing the nascent antibody chains to the periplasm (the region between the inner and outer membranes) of the bacterium where the intradomain disulphide bridge is formed and the V_H and the V_L pair. Periplasmic expression in the bacterium mimics the natural production route in the endoplasmic reticulum of the lymphocyte. In the natural biosynthetic process, the heavy and light antibody chains both carry a signal peptide which is cleaved during transport from the cytoplasm over the membrane into endoplasmic reticulum where the disulphide bridges are formed. To achieve secretion into the periplasmic space, the antibody fragment sequence is fused with a bacterial signal sequence.

Antibody fragments can also be expressed on the surface of filamentous bacteriophage (viruses which infect bacteria) by fusing them to a coat protein (phage display, Figure 3.8). The phage particles are long and thin and are essentially the genome, a circular single-stranded DNA molecule, covered with thousands of copies of a small coat protein (pVIII) and, at one end of the particle, by five copies of the coat protein pIII. In phage assembly, the coat proteins are exported into the bacterial periplasm and antibody V_H and V_L chains can be similarly directed into the bacterial periplasm by incorporation of a signal sequence. The gene encoding a scFv antibody fragment is inserted into the 5'-end of the gene coding for coat protein III and expression of the fusion product results in the antibody fragment being presented on the phage surface while its genetic material resides within the phage particle. Three copies of the fusion protein are usually displayed. The display of antibody fragments on phage means that the phage can be treated as if it were an antibody and thus be selected for the desired binding affinity.

Phage display has the potential to mimic the antibody arm of the immune system. The natural antibody system is fundamentally combinatorial with

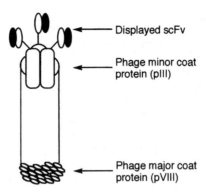

Figure 3.8   Display of scFv on the surface of filamentous phage. A gene encoding the scFv is cloned with the phage minor-coat protein gene such that the functional fusion protein is displayed in the surface of the phage particle

diversity being generated by random combination of gene fragments (Scheme 3.15). Individual combinations of different gene segments results in an estimated 100 million (10^8) different B cells circulating in the bloodstream, each coding for a different antibody. Each specific antibody from the repertoire is presented on the surface of its B cell and on contact with specific antigen the B cell is activated and begins to proliferate and mature into an antibody-secreting plasma cell. Displaying protein molecules on the surface of phage can imitate the linkage outside the B cell and by creating a large repertoire or library of antibodies the natural combinatorial process can be mimicked.

The genetic manipulations needed to produce a library of antibody fragments are not unlike the initial steps in the creation of chimeric and humanized antibodies. In all of these procedures the first step is PCR amplification of V_H and V_L genes. Whereas in the generation of humanized antibodies the first PCR template is mRNA extracted from a defined hybridoma cell line, the starting material for a phage display library is mRNA from a pool of B cells (Scheme 3.16). The mRNA found in a mixture of B cells contains all of the information for the complete set of antibody genes. The cDNA obtained from B-cell mRNA is, as in the humanized protocols, further amplified in a second round of PCR by using primers complementary to the termini of the variable genes. The scFv repertoire is then created by combining the amplified V_H and V_L genes in a PCR reaction containing linker DNA. The linker DNA codes for the peptide $(Gly_4Ser)_3$ and has regions of homology with the 3′ end of the amplified V_H gene and 5′ end of the V_L gene. Alternatively, if oligonucleotide primers which code for the peptide linker are used in the second PCR reaction mixture then the two V regions can be assembled into a single piece of DNA directly. The resulting scFv gene repertoire is re-amplified with primers containing restriction sites to permit cloning into a phage vector. The scFv gene repertoires are digested with restriction enzymes and ligated into a phage

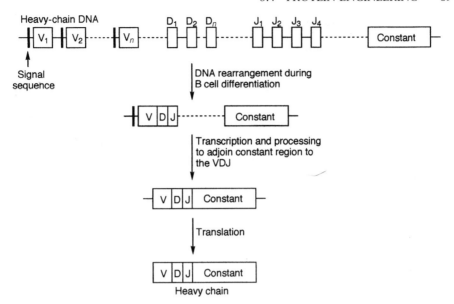

Scheme 3.15 Generation of antibody diversity. DNA encoding antibody heavy and light chains are found on different chromosomes. A large number of gene segments encode the variable region of an antibody chain. During the natural development of the B-cell repertoire in the body, each B cell develops the gene for its specific antibody by recombination of variable (V), joining (J) and, diversity (D) sequences. Each light-chain variable region is encoded by a DNA sequence assembled from a V and a D segment. whereas heavy chains are assembled from V, J, and D segments (shown). The assembly of different combinations of V, J, and D gene segments produces a large number of V_H and V_L regions which make up the diversity of antigen-binding sites. The information for the complete set of antibody genes is found in a mixture of B cells

display vector which has been digested with the same restriction enzymes. The vector also contains the gene encoding the coat protein pIII. Expression is under the control of the *lac* promoter and the pelB signal sequence directs expressed protein to the periplasm. The vector ligation mixes are introduced into *E. coli* and a phage antibody library is produced (Scheme 3.17).

Since the antibody fragments on the surface of the phage are functional, phage bearing antigen-binding antibody fragments can be separated from non-binding phage by antigen affinity chromatography. In this, antigen is immobilized either on a column matrix or on the surface of a plastic dish or tube. In the method known as panning, the phage library is first incubated in an antigen-coated dish. The dish is then washed extensively to remove non-specific phage before elution of antigen-specific phage by using either a mild denaturant or soluble antigen. As the genetic material encoding the antibody fragment resides within the phage particle, bacteria can be infected with the eluted phage and more phage can be grown with identical binding characteristics and subjected to another round of selection. The selection procedure enriches binders from the phage library and in order to select those phage

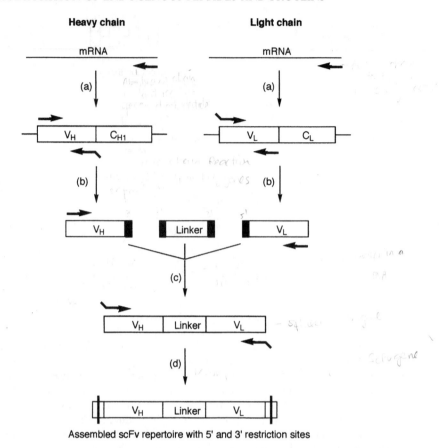

Assembled scFv repertoire with 5' and 3' restriction sites

Scheme 3.16 Creation of human scFv gene repertoires using PCR. (a) In separate reactions, mRNA encoding V_H and V_L genes is primed with antibody heavy- and light-chain constant region-specific oligonucleotides (represented by bold arrows) and cDNA is synthesized. (b) cDNA is used in a second round of PCR to amplify V_H and V_L genes separately. (c) The amplified V_H and V_L genes are spliced together in a PCR reaction mixture containing linker DNA (the bold regions represent sequences that are homologous between the 5' end of the linker and the 3' end of V_H and the 3' end of the linker and the 5' end of V_L, respectively). (d) The spliced scFv gene repertoire is re-amplified with flanking primers containing appended restriction sites

antibodies with the best affinities, decreasing amounts of antigen are used in each round. Each cycle of enrichment can increase the proportion of phage antibodies within the population by a factor up to 1000-fold and so in two rounds 1 000 000-fold enrichment is possible. Ultimately, the phage population will become essentially clonal. Usually after two-to-four rounds of selection, phage antibody clones can be assayed directly for their ability to bind specific antigens by immunoassay techniques such as ELISA.

There are two types of libraries, namely 'naive' and immunized. In immune libraries, V-genes are derived from mRNA of B cells of an immune source e.g.

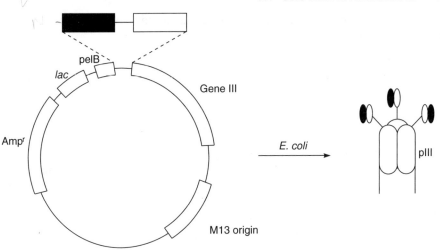

Scheme 3.17 Generation of scFv antibody fragment library displayed on phage. The scFv repertoire is digested with appropriate restriction enzymes and ligated into a phage display vector. A bacterial signal sequence (pelB) is fused to the N-terminus of the protein to cause secretion through the inner membrane of the *E. coli* cell. The ligation mix is used to transform *E. coli* and a phage antibody library is produced. Antibody genes are depicted by a dark shaded bar (heavy chain) and a light shaded bar (light chain); *lac*, *lac* promoter, Amp^r, ampicillin resistance. Antibody fragments are depicted as dark (heavy chain) or light (light chain) shaded ovals

immunized animals or, in some instances, human immune B cells. If the source of the genetic material comes from an unimmunized donor then a 'naive' immune phage antibody will be made. Antibody fragments against any chosen antigen may be selected from both types of library but the affinities for a given antigen are lower in naive libraries. Identifying and retrieving a scFv to an antigen from a naive library is much harder than from an immune library enriched in antigen-specific antibody fragments and a very good screening system is required. Currently, scFvs selected from immune and naive libraries are extremely useful for research purposes but their affinity is not sufficient for many therapeutic applications in immunotherapy or for use in sensitive diagnostics.

It is the rapid sorting as a result of the linkage of recognition and replication which makes phage display technology such a powerful and important tool. Phage-displayed antibodies are immediately accessible to further genetic manipulation and they can be reformatted as whole antibodies for expression in eukaryotic cells. Naive antibody libraries bypass the need for immunization and hybridoma technology and offer an *in vitro* route to antibodies which are 100% human. A drawback of these methods is that the construction of libraries is technically difficult although it is not unlikely that within a few years these may be commercially available. If the selection and efficiency can be improved, phage libraries will probably come to complement hybridoma technology as a means of producing efficacious human antibodies suitable for therapy.

Further Reading

Textbook and review articles

- J. R. Adair and T. P. Wallace, Site-Directed Mutagenesis, in *Molecular Biomethods Handbook*, R. Rapley and J. M. Walker (Eds), Humana Press, Totowa, NJ, 1998, pp. 347–360.
- C. A. K. Borrebaeck (Ed.), *Antibody Engineering*, 2nd Edn, Oxford University Press, New York, 1995.
- F. Breitling and S. Dübel, *Recombinant Antibodies*, Wiley, New York, 1999.
- J. R. Crowther, Enzyme-Linked Immunosorbent Assay (ELISA), in *Molecular Biomethods Handbook*, R. Rapley and J. M. Walker (Eds), Humana Press, Totowa, NJ, 1998, pp. 595–617.
- A. S. Dutta, Design and Therapeutic Potential of Peptides, *Adv. Drug Res.*, 1991, **21**, 145–286.
- R. Edwards (Ed.), *Immunodiagnostics: A Practical Approach*, Oxford University Press, Oxford, UK, 1999.
- J. Gante, Peptidomimetics – Tailored Enzyme Inhibitors, *Angew. Chem. Int. Ed. Engl.*, 1994, **33**, 1699–1720.
- A. Giannis and T. Kolter, Peptidomimetics for Receptor Ligands – Discovery, Development, and Medical Perspectives, *Angew. Chem. Int. Ed. Engl.*, 1993, **32**, 1244–1267.
- M. Goodman and S. Ro, Peptidomimetics for Drug Design, in *Burger's Medicinal Chemistry and Drug Discovery*, 5th Edn, Vol. 1. M. E. Wolff (Ed.), Wiley, New York, 1995, pp. 803–861.
- H. de Haard, P. Henderikx and H. R. Hoogenboom, Creating and Engineering Human Antibodies for Immunotherapy, *Adv. Drug. Del. Rev.*, 1998, **31**, 5–31.
- G. T. Hermanson, *Bioconjugate Techniques*, Academic Press, San Diego, CA, 1996.
- T. Kieber-Emmons, R. Murali and M. I. Greene, Therapeutic Peptides and Peptidomimetics, *Curr. Opin. Biotechnol.*, 1997, **8**, 435–441.
- R. J. Kreitman, Immunotoxins in Cancer Therapy, *Curr. Opin. Immunol.*, 1999, **11**, 570–578.
- J. Rizo and L. M. Gierasch, Constrained Peptides: Models of Bioactive Peptides and Protein Substructures, *Annu. Rev. Biochem.*, 1992, **61**, 387–418.
- R. A. G. Smith, J. M. Dewdney, R. Fears and G. Poste, Chemical Derivatization of Therapeutic Proteins, *Trends Biotechnol.*, 1993, **11**, 397–403.

- J. E. Thompson and A. J. Williams, Phage-Display Libraries, in *Molecular Biomethods Handbook*, R. Rapley and J. M. Walker (Eds), Humana Press, Totowa, NJ, 1998, pp. 581–594.
- T. J. Vaughan, J. K. Osbourn and P. R. Tempest, Human Antibodies by Design, *Nat. Biotechnol.*, 1998, **16**, 535–539.

Research publications

- W. Bauer, U. Briner, W. Doepfner, R. Haller, R. Huguenin, P. Marbach, T. J. Petcher and J. Pless, SMS 201-995: A Very Potent and Selective Octapeptide Analogue of Somatostatin with Prolonged Action, *Life Sci.*, 1982, **31**, 1133–1140.
- P. Carter, L. Presta, C. M. Gorman, J. B. B. Ridgway, D. Henner, W. L. T. Wong, A. M. Rowland, C. Kotts, M. E. Carver and H. M. Shepard, Humanization of an Anti-p185[HER2] Antibody for Human Cancer Therapy, *Proc. Natl. Acad. Sci. USA*, 1992, **89**, 4285–4289.
- A. S. Dutta, B. J. A. Furr, M. B. Giles and B. Valcaccia, Synthesis and Biological Activity of Highly Active α-Aza Analogues of Luliberin, *J. Med. Chem.*, 1978, **21**, 1018–1024.
- A. Heppeler, S. Froidevaux, H. R. Mäcke, E. Jermann, M. Béhé, P. Powell and M. Hennig, Radiometal-Labelled Macrocyclic Chelator-Derivatised Somatostatin Analogue with Superb Tumour-Targeting Properties and Potential for Receptor-Mediated Internal Radiotherapy, *Chem. Eur. J.*, 1999, **5**, 1974–1981.
- L. Reichmann, M. Clark, H. Waldmann and G. Winter, Reshaping Human Antibodies for Therapy, *Nature (London)*, 1988, **332**, 323–327.
- R. A. G. Smith, R. J. Dupe, P. D. English and J. Green, Fibrinolysis with Acyl-enzymes: A New Approach to Thrombolytic Therapy, *Nature (London)*, 1981, **290**, 505–508.

4

The Immune System

4.1 Overview

The immune system is the means by which the body defends and heals itself; it initiates the destruction and elimination of invading organisms and any toxic molecules produced by them. Immune reactions are consequently destructive and should only be made in response to material which is foreign to the host and not to that of the host itself. If the latter situation occurs and the immune system reacts destructively against the host's own molecules then autoimmune diseases such as rheumatic fever and multiple sclerosis develop. There are two branches to the immune system (Scheme 4.1) and often both are employed against infections. The B cells dominate one part of the system and events associated with their activation are referred to as 'antibody-mediated' response. The T cells dominate the other branch and when they are activated this is referred to as a 'cell-mediated' response. Some diseases may be treated by immunotherapy in which the immune system is manipulated by modification of these endogenous responses.

4.2 The Antibody-Mediated Response

4.2.1 Polyclonal antibodies

The antibody response involves production of immunoglobulins by the white blood cells (B lymphocytes) in the bone marrow. These proteins circulate in the blood stream permeating other body fluids where they bind specifically to the foreign substance or antigen (*anti*body *gene*rator) that induced them. Effector molecules then recognize and bind to the antibody bound to the antigen and trigger its elimination.

Once an antibody has been made, an individual can produce relatively large quantities of it very easily – this is the 'memory' of the immune system. The

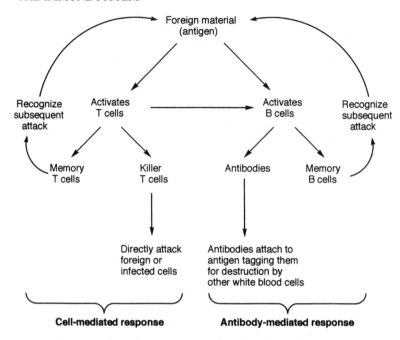

Scheme 4.1   The immune mechanism. Reproduced with modifications from Grace, E.S., *Biotechnology Unzipped: Promises and Realities*, 1997, by permission of the Joseph Henry Press.

next time the same antigen is encountered the immune response is faster and stronger.

All antibodies have the same basic structure – they are Y-shaped proteins of approximately 150 kDa that consist of two identical arms and a tail (Figure 4.1). The two arms of the antibody are involved in binding to the antigen and are referred to as Fab (fragment antibody binding) while the tail, the Fc (fragment crystalline) unit, links the antibody to effector molecules. The antibody molecule is made up of four polypeptide chains, two identical heavy (H) chains of approximately 450–575 amino acids which span the Fab and Fc units and two identical light (L) chains of approximately 200 residues which are associated only with the Fab unit. These chains are covalently linked by interchain disulphide bonds in such a way that the antibody molecule consists of two identical halves. Both H and L chains contain variable (V) and constant (C) regions and the antigen-binding site is formed by the spatial juxtaposition of six polypeptide segments known as the complementarity determining regions (CDRs) which are found near the tip of the Y arms in the variable region. These polypeptide segments, three from each of the H and L chains, adopt loop structures and variations in their length and amino acid sequence are primarily responsible for the differences in specificity and affinity of different antibodies (Figure 4.2). The parts of the antigen that combine with the antigen-binding site on an antibody molecule are known as

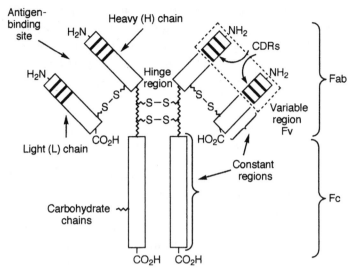

Figure 4.1 An outline of a typical antibody molecule illustrating the four-chain structure (2H + 2L). Each antibody L or H chain is structurally segmented into regions that are either of the variable type or the constant type. The Fab fragment comprises the Fv (variable domains of the L and H chains) and part of the constant region, while the Fc fragment comprises the two C-terminal halves of the H chains. The CDRs are shown in bold.

antigenic determinants or epitopes. Antibodies are tailored to fit the epitope through an ingenious combinatorial diversification strategy whereby a relatively small number of gene segments that code for the variable regions of the H and L chains can be recombined to produce millions of different molecules (see Scheme 3.15). This enables the immune system to develop a highly focused response to any given infection.

The antigen–antibody complex activates the complement system, a cascade of serum enzymes, which results in lysis (rupture) of the cell membrane possessing the epitope and leads to cell destruction. The cascade process is similar to that which causes blood to clot; proenzymes are sequentially activated in an amplifying series of proteolytic reactions. As in blood clotting,

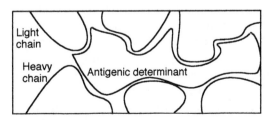

Figure 4.2 Schematic representation of how the three variable regions in each of the H and L chains together form the antigen-binding site of an antibody molecule. The specificity of the antibody–antigen interaction is based on the fitting together of two molecules of complementary shape

this allows the body to respond to trauma rapidly and effectively. Foreign cells coated with antibodies can also be destroyed by a process known as phagocytosis. A macrophage, one of a class of white blood cells, engulfs the foreign cell and then destroys it internally. The mechanism, known as antibody-dependent cellular cytotoxicity, is triggered by the Fc parts of the bound antibody molecule interacting with receptors on the surface of the macrophage.

Most antigens have a variety of antigenic determinants and consequently when the immune system encounters such species a heterogeneous mixture of antibodies is produced. These antibodies are known as polyclonal antibodies.

Polyclonal antibody preparations can be administered clinically to give immediate immunity in situations where an individual may not have an adequate defence against a particular pathogen or other harmful antigenic substance. Immunity persists for two to three weeks and is largely an emergency procedure. As the protective antibodies are not actively generated by the body's own immune system, the use of such preparations is referred to as passive immunization.

Long-lasting defence against an infectious agent is achieved by the process of active immunization (vaccination) whereby an individual's own immune system launches an immunological response to an antigenic substance. Active immunization with the appropriate vaccine frequently accompanies or follows passive immunization. For example, rabies immunoglobulin is used in conjunction with active immunization employing rabies vaccine as part of the post-exposure treatment for prevention of rabies in persons who have received bites from rabid animals or animals suspected of being rabid. Tetanus immunoglobulin neutralizes the toxin formed by *Clostridium tetani* and provides temporary immunity against tetanus and is used as part of the management after injury in persons unimmunized or incompletely immunized against tetanus. These specific immunoglobulins preparations are obtained from the plasma or serum of blood donors who have been immunized with a particular vaccine or who are recovering from the infection in question and have developed a high titre of antibodies. Polyclonal antibody preparations may alternatively be obtained from the blood of healthy animals, usually the horse, that have been immunized against the appropriate pathogenic substance. Antibody preparations isolated from animals are often referred to as antisera to distinguish them from immunoglobulins which are antibody preparations obtained from human sources. Antisera are normally used for passive immunization against the poisonous antigens present in the venom of some snakes, scorpions and spiders. Adverse reactions are liable to occur after the administration of any serum of animal origin. Serum sickness may occur 7 to 10 days after the injection of the antisera and occurs when antibodies, produced in response to antigen stimulation, react with the antigen to form circulating soluble immune complexes (see also HAMA response in the following Section 4.2.2). Severe allergic reactions resulting in anaphylactic shock in which the outstanding feature is a sudden general circulatory collapse can be life-threatening. Because of these adverse reactions, immunoglobulins which are of human rather than of animal origin are preferred as passive immunizing agents.

Routine blood donations contain many antibodies to infectious diseases prevalent in the general population. Typical antibodies present include those against hepatitis A, measles, mumps, rubella, diphtheria, polio and chicken pox and normal immunoglobulin may therefore be used to provide passive immunization against such diseases. For example, the likelihood of a clinical attack of rubella in unimmunized pregnant women exposed to rubella may be reduced if the normal immunoglobulin is given soon after exposure.

As with other proteins isolated from human material, strenuous efforts are made to screen the donor material to prevent the transmission of infection such as hepatitis B and HIV.

4.2.2 *Monoclonal antibodies*

Monoclonal antibodies (mAbs) are monospecific, i.e. all of the molecules in a particular preparation are identical. These homogeneous preparations are manufactured by hybridoma technology which allows unlimited quantities of an antibody with a particular specificity to be obtained (Scheme 4.2). The procedure was developed in the 1970s by Milstein and Köhler and involves fusing immortal, non-antibody secreting, myeloma cells with antibody-producing B lymphocytes from the spleen of a mouse which has been immunized against the antigen of interest. The resulting fused cells, the hybridomas, retain the immortal characteristics of the myeloma cells while secreting large quantities of monospecific antibodies. Each hybridoma produced from the fusion protocol makes an antibody with specificity to a particular epitope of the antibody-eliciting molecule but most hybridomas will secrete antibodies directed against those antigens which are not of interest. The trick in producing mAbs is to isolate and clone single hybridomas so that only one immunoglobulin molecule will be secreted. This is a painstaking process of screening, cloning and rescreening, but once the correct hybridoma cell line has been obtained an unlimited quantity of supernatant containing antibody can be collected.

The constant availability of a well characterized, pure and homogeneous material with high specificity has made mAbs valuable agents in the diagnosis and treatment of disease. The first successful clinical use of mAbs was to suppress rejection of kidney transplants. A major problem in organ transplantation is rejection of donor tissue by the recipient; the host's immune system sees the donor organ as foreign and mounts an immune response. This produces the clinical syndrome of graft-vs-host disease which can be lethal. Rejection of transplanted tissue is mediated by T cells, the vehicles that regulate cellular immunity, and form the second branch of the immune

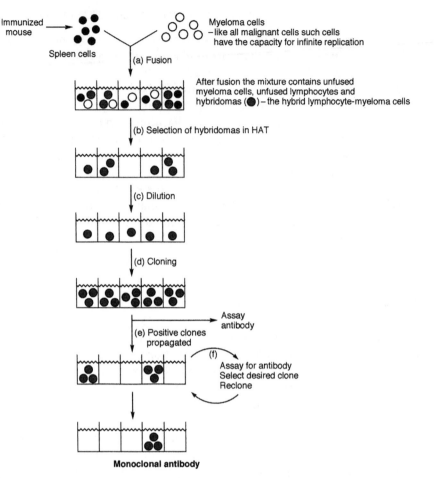

Monoclonal antibody

Scheme 4.2 Preparation of a monoclonal antibody. (a) Spleen lymphocytes from a mouse immunized with the antigen of interest are mixed with hypoxanthine phosphoribosyl transferase (HPRT) negative myeloma cells in the presence of poly(ethylene glycol) (PEG). HPRT is a key enzyme in the purine metabolism pathway and when HPRT negative cells are grown in a mixture of hypoxanthine, aminopterin and thymidine (HAT medium) the cells die because they can no longer synthesize DNA. This means that myeloma HPRT negative cells that do not fuse can be easily destroyed after the fusion process. In the presence of PEG, the membranes of the myeloma and spleen lymphocyte cells can fuse, thus allowing the nucleus of one cell to enter the cytoplasm of another. (b) After the fusion process, the hybridomas are selected from the unfused cells. This is done by plating the cells out on to tissue culture dishes and changing the composition of the tissue culture medium to HAT. Only hybridomas will grow successfully in the HAT medium and thus can be selected from unfused cells; normal B lymphocytes from the immunized animal's spleen cannot grow in tissue culture for more than a few days and therefore die naturally, while unfused HPRT myeloma cells can also not grow in the HAT culture medium. (c) The hybridomas are diluted and distributed into wells such that statistically each well has on average only a single cell. (d) The cells are grown for 14 days to produce antibody and after this period the supernatants are screened by using an ELISA or radioimmunoassay. In this way, those wells containing antibodies which recognize the antigen of interest are identified. (e) The supernatants from the positive wells are rediluted, grown up and rescreened, and (f) this process is repeated until monoclonality is obtained

Table 4.1   Monoclonal antibodies marketed for therapeutic purposes. Note: Campath-1, a murine antibody directed against the surface antigen CDw52 which is expressed in lymphocytes, was the first antibody to be reshaped by genetic engineering. The humanized form, Campath-1H, has been evaluated as a treatment of blood malignancies and for some autoimmune disorders although the results were variable

Antibody name(s)	Antibody type	Target antigen	Indication	Year product licence granted
Muromonab-CD3 (Orthoclone OKT3)	Murine	CD3	Allograft rejection	1986
Infliximab (Remicade)	Chimeric	TNF-α	Rheumatoid arthritis	1998
			Crohn's disease	1999
Basiliximab (Simulect)	Chimeric	CD25	Allograft rejection	1998
Daclizumab (Zenapax)	Humanized	CD25	Allograft rejection	1997
Rituximab (Rituxan)	Chimeric	CD20	Non-Hodgkin's B-cell lymphoma	1997
Trastuzumab (Herceptin)	Humanized	HER2/neu	Metastatic breast cancer	1998

response (see Section 4.4 below). An anti-T-cell mAb known as muro-monab-CD3 (Table 4.1) specifically targets the accessory glycoprotein CD3 expressed on the surface of T cells. This CD3 antigen is essential to antigen recognition and responses and so binding of muromonab-CD3 specifically blocks T-cell generation and function to exert a profound immunosuppressive effect. Muromonab-CD3 was marketed in 1986 for the treatment of acute allograft rejection in organ transplant patients. However, after an initial course of treatment most patients develop human anti-mouse antibodies (HAMA) which block the muromonab-CD3 binding site and over time this neutralizes the murine antibody. Successful use of muromonab-CD3 for suppression of graft rejection relies upon single use or multiple use within a short period of a week to 10 days before the anti-antibody response develops. Many treatment protocols following organ transplant use a variety of immunosuppressive agents to enhance efficacy and minimize toxicity. While muromonab-CD3 may be effective as an initial treatment to prevent rejection, immunosuppression is frequently maintained by using a triple drug regime consisting of a calcineurin inhibitor (cyclosporin A, or tacrolimus, see Section 4.2.2 below), corticosteroids and an antimetabolite such as azathioprine.

The HAMA response, a short half-life in serum and poor recognition of rodent constant regions by human effector functions has limited the therapeutic effectiveness of mouse mAbs. Therefore, a human or nearly human antibody is needed for clinical application. Production of human mAbs is problematic; human myelomas are difficult to grow in tissue culture and it is necessary to rely either on fortuitous immunizations for obtaining precursor cells of the appropriate antigen specificities or to use the natural repertoire of antibody specificities. Furthermore, the only practical source of cells for generating antibodies is peripheral blood. An alternative strategy uniquely available to the human system is through the Epstein–Barr virus which can transform and immortalize human B lymphocytes. The drawback of this technique is that the transformants seldom secrete large amounts of antibody.

An alternative approach to overcome the innate immunogenicity of murine mAbs in humans is to use genetic engineering techniques and convert the mouse antibodies into mouse-human hybrids (Figure 4.3). The first generation of hybrids were chimeric mAbs with murine variable regions and human constant regions. Infliximab is an anti-tumour necrosis factor α (TNF-α) chimeric mAb that was licensed in 1998 for Crohn's disease and in 1999 for rheumatoid arthritis. Crohn's disease is an inflammatory bowel disorder and infliximab blocks the action of the pro-inflammatory cytokine TNF-α, thus producing clinical improvement. TNF-α and other cytokines have a possible role in the pathogenesis of rheumatoid arthritis, a common, progressive and crippling inflammatory disease. There is no curative treatment for rheumatoid arthritis and management of the disease is aimed at alleviating pain and maintaining joint function. Many drugs have been tried in rheumatoid arthritis and significant improvement in joint pain, inflammation and mobility was found in patients treated with infliximab either alone or in combination with methotrexate.

Chimeric mAbs can be produced by direct chemical coupling of Fab fragments through thiol–ester linkages to human Fc prepared by enzymatic digestion or by ligating genetically cloned mouse variable sequences to human constant regions by rDNA techniques (see Section 3.4.3 earlier). The presence of mouse variable regions is however sufficient to trigger an anti-mouse antibody response in patients, resulting in decreased efficacy of the therapeutic antibody. To overcome this problem, more human-like antibodies are produced by grafting the six CDRs from the heavy and light chains of a rodent antibody on to a human framework. These 'humanized' antibodies only contain approximately 5 to 10% of murine residues and are therefore less

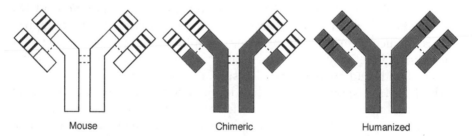

Mouse   Chimeric   Humanized

Figure 4.3 Structures of mouse-human chimeric and humanized monoclonal antibodies. The regions in white represent mouse sequences and the shaded regions represent human sequences. Mouse CDRs are shown in bold. The chimeric antibody consists of human constant sequences with mouse variable domains, while the humanized form is composed of the original mouse derived CDRs within a human immunoglobulin framework

likely to produce the HAMA effect. The humanization procedure (see Section 3.4.3 earlier) requires complicated manipulations at the genetic level.

The humanized mAb daclizumab is an immunosuppressant agent that reduces acute rejection in renal transplant recipients. It is specific for the α-sub-unit (Tac/CD25) of the interleukin 2 receptor (IL-2R) on activated T cells and achieves immunosuppression by competitive antagonism of IL-2R-induced T-cell proliferation. Daclizumab and basiliximab, a chimeric anti-IL-2R mAb, have advantages over the murine mAb to the IL-2R, including improved effector function, a low potential for immunogenicity and a long half-life. When added to standard cyclosporin-based immunosuppressant therapy, daclizumab significantly reduces the frequency of acute rejection of renal transplants. As chimeric and humanized antibodies contain the human Fc region they can recruit complement and human effector systems and so can be effective in cancer therapy where cell destruction is required. An anti-CD20 chimeric mAb (rituximab) and an anti-HER2/neu humanized mAb (trastuzamab) have been licensed for the treatment of lymphoma and breast cancer, respectively (see also Section 4.5.1 below).

Other attempts to limit the HAMA response have included the use of antibody fragments such as Fab', F(ab')$_2$ and single-chain antibodies (Figure 4.4). Biochemical digestion procedures are used to prepare the Fab' and F(ab')$_2$ fragments, while the single-chain species are obtained by application of genetic engineering techniques (see Section 3.4.4 above). Single-chain molecules (scFvs) are designed to consist of variable light and variable heavy domains of an antibody tethered together by a peptide linker such that the antigen-binding site is regenerated in a single polypeptide. Although

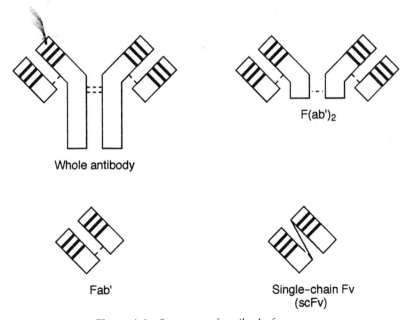

Whole antibody

F(ab')$_2$

Fab'

Single-chain Fv (scFv)

Figure 4.4   Structure of antibody fragments

fragments of murine antibodies are less immunogenic in humans than the whole antibody, these proteins are rapidly cleared from the blood and have decreased antigen binding avidity and decreased molecular stability compared with the parent compound.

A variety of diseases produce abnormalities in the blood, urine and cerebrospinal fluid and detecting and measuring such anomalies can be vital to the diagnosis of particular conditions. The specificity of mAbs makes them ideal reagents for detecting the presence of individual compounds in bioassays. One particular area where mAbs has had a significant impact is in the diagnosis and treatment of cancer. Some tumours secrete specific marker substances which, if detected, may be evidence of malignant disease. For example, carcinoembryonic antigen (CEA), a complex glycoprotein that occurs on the surface of most colon cancers, is shed by the cancer cells into the blood. Detecting CEA in a blood sample may assist in the diagnosis of a tumour in a patient. Furthermore, monitoring the amount of the marker in serum is useful in the follow-up to treatment as a rise in its level may indicate the recurrence of a malignancy in a particular patient. Elevated levels of human chorionic gonadotrophins (hCGs) are indicative of the presence of choriocarcinoma and a decrease in hCGs correlates well with successful therapy.

Monoclonal antibodies to tumour-associated antigens (TAAs) can also be used to locate malignant cells within patients suffering from certain types of cancer. A radioisotope, usually a low-energy γ-emitter such as ^{111}In, ^{131}I, or ^{99m}Tc, is coupled to the antibody, the antibody is taken up by the tumour and the localized radioisotope is then detected by an external radioimaging device. Iodine-131 has a high-energy β-emission and a weak associated γ-emission which makes it a versatile radioisotope as it can be used for simultaneous imaging and therapy (see below). Antibodies can be directly labelled with radioactive iodine by the chloramine T procedure; Na^{131}I is oxidized with chloramine T to the mixed halogen species ^{131}ICl, which then halogenates tyrosyl and histidyl side-chains (see Section 3.3 earlier). Other radioisotope–antibody conjugates are formed by indirect labelling; a chelating group is first attached to the antibody and the radioisotope is conjugated to the antibody by binding at these specific attached chelation groups. In common with many protein modification techniques, these radiolabelling procedures will modify all accessible side-chains of a particular type of amino acid and if critical residues are altered then the immunological activity of the antibody may be affected. Diagnostic imaging is not yet a routine investigative procedure but useful findings are emerging from patient studies. Imaging patients with colon and ovarian carcinomas with ^{111}In-Oncoscint, an antibody specific for an antigen found on most colon and ovarian cancer cells, showed that a positive scan is a reliable predictor of cancer but a negative scan is unreliable and further testing is therefore required.

MAbs can carry not only diagnostic isotopes but also therapeutic agents such as antitumour drugs, toxins or high-activity isotopes. The rationale behind this approach to cancer immunotherapy is to target the toxic substance directly to tumour cells and spare normal cells (Scheme 4.3). The antibody provides the specificity, while the toxic agent acts as a potent warhead – the ultimate in 'magic bullets', a term originally proposed by Paul Ehrlich, the forefather of modern chemotherapy. As with all mAb therapies, there are difficulties in penetrating tumour masses and the technique is proving to be more successful with blood malignancies rather than with solid tumours such as those affecting the colon, breast or lung. Coupling the antibody and toxic substance without abrogating the activity of both molecules and forming a stable conjugate are similar problems to those encountered in diagnostic imaging. A more stringent requirement demanded of targeted immunotherapy is that the specificity of the antibody must be absolute – the toxin must be delivered only to the tumour and not to normal tissues.

For targeted delivery to be successful it is necessary for the cytotoxic agent to be extremely active. Ricin, a plant toxin, is one of the most cytotoxic substances known and acts by catalytically inhibiting protein synthesis. Only a few molecules of ricin therefore need to reach the cytoplasm in order to kill the target cell. Ricin consists of two sub-units, i.e. the enzymatically active A chain

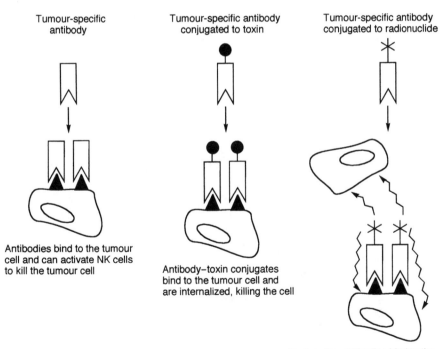

Scheme 4.3   Cancer therapy using antibodies. Radiation and toxins are directly cytotoxic to targeted cells, although radiation can also have the benefit of a bystander effect

disulphide bonded to the targeting B chain. In a ricin conjugate, the B chain of ricin is removed to prevent the toxin from binding to normal cells and then the A chain is linked to the antibody by formation of a disulphide bond. The A chain is obtained either by reducing whole ricin or prepared, without glycosylation, by recombinant techniques in *E. coli*. Zolimomab aritox, a murine mAb against the CD5 antigen of lymphocytes conjugated with the ricin A chain, has been used experimentally for the prophylaxis of graft-vs-host disease in patients receiving bone-marrow transplants.

Calicheamicins are members of the enediyne family of antibiotics which are among the most toxic antitumour compounds known. They exert their toxicity by producing double-stranded breaks in DNA which prevents cell division and thence cell replication. Immunoconjugates have been prepared by reacting a hydrazide of the most potent and abundant member of the enediynes, calicheamicin γ_1 (**4.1**), with oxidized sugars on the constant region of the antibody. The hydrazide derivatives are prepared by displacement at the methyltrisulphide moiety with mercaptohydrazides. For example, reaction of calicheamicin γ_1 with 3-mercaptopropionyl hydrazide ($NH_2-NH-CO-CH_2-CH_2-SH$) gives the simple derivative **4.2**. Although not an established chemotherapeutic agent, an immunoconjugate consisting of calicheamicin γ_1 bound to a humanized anti-CD33 mAb has been used in early clinical trials of patients with acute myeloid leukaemia with encouraging results.

(**4.1**) R = SCH₃, Calicheamicin γ_1

(**4.2**) R = $-CH_2-CH_2-C(=O)-NHNH_2$

Administration of radiolabelled mAbs is another method of selectively delivering cytotoxic agents. Radioisotopes emit particles capable of inducing lethal DNA damage to cells. Radiation has the power to kill cells within a given range and so radiolabelled antibodies not only kill tumour cells bearing antigen but also antigen-negative tumour variants or tumour cells not reached by the mAbs (bystander effect). Therapy with radioimmunoconjugates is most frequently conducted with the high-energy β-emitting radionuclides ^{131}I (half-life 193 h) and ^{90}Y (half-life 64 h) and has been most effective for haematological malignancies. For example, anti-Tac, a mAb directed against the IL-2R CD25 antigen, is a promising agent for the treatment of adult T-cell leukaemia. IL-2R is found on activated but not resting T cells and is over-expressed in adult T-cell leukaemia. In parallel studies, one group of patients was treated with anti-Tac coupled to ^{90}Y, while the second group received the unlabelled antibody. Both treatments gave at least partial remission in most cases but there were more complete remissions with the radiotherapy and at a lower dose of the antibody.

MAb-based treatments for human diseases are on the increase. Genetic engineering techniques allow for the production of chimeric and humanized mAbs which, with minimal immunogenicity, long half-lives and efficient interaction with human effectors, are becoming established as blocking reagents for a variety of immune and infectious diseases. Naked mAbs by themselves may not be sufficiently toxic to kill cancer cells but the use of mAbs to deliver cytotoxic agents such as radioisotopes specifically to tumours has become a promising approach for treating the latter.

4.3   Vaccines

A vaccine is used as a prophylactic measure to prevent the future development of a specific disease. Vaccination involves the administration of an appropriate antigen which stimulates the immune system to generate its own immunological response and the process is therefore known as active immunization. As well as inducing B lymphocytes to produce protective antibodies which recognize and bind to the antigen, the cellular component of the immune system involving T lymphocytes may also be activated. After the immune system has successfully eliminated the antigenic material, long-lived B and T lymphocytes, known as memory cells, remain in circulation. These memory cells allow the body to mount a faster and stronger immunological response if the same or antigenically similar pathogen is subsequently encountered. The pathogen is thereby destroyed before it can establish an infection. The process of vaccination is thus designed to exploit the natural defence mechanism conferred by the immune system and the vaccine works by mimicking the natural infection process without inducing concomitant disease.

Vaccination programmes have achieved the global eradication of smallpox and have eliminated or substantially reduced morbidity and mortality for a wide range of other diseases. The vaccine itself is a preparation of dead or attenuated live virus or bacteria or antigens derived from these sources. The disadvantages of such vaccines is that not only is it necessary to ensure that the organisms are killed or have low pathogenicity but these preparations may be significantly less immunogenic than their wild-type counterpart. A rigorous inactivation procedure (typically treatment with a strong denaturing disinfectant such as formaldehyde or phenol) is essential to kill the pathogen, although the harsh conditions makes the vaccine less effective. The process causes denaturation of the proteins and carbohydrates that are essential for the organism to live and infect a host, but if treated properly the surface antigens are left intact and the preparation is recognized as the original antigen. The Salk polio and influenza vaccines are inactivated virus vaccines and both have been used successfully for many years. With vaccines based on live attenuated strains of viruses and bacteria there is a risk, albeit small, that they could revert to the pathogenic state. Furthermore vaccines from live organisms can elicit severe infections in recipients whose immune responses are deficient, e.g. persons with congenital or acquired immunodeficiency. An attenuated or weakened strain of a pathogen is obtained by passage through many generations of host animals. The idea being that the animal and pathogen, if both are to survive, need to adapt to live with each other without either being killed. Polio virus attenuated in this fashion in monkey tissue is the basis of the oral Sabin polio vaccine. Vaccines against measles, mumps and rubella are also composed of attenuated live viruses. The measles, mumps and rubella vaccines can be combined into one injection (the MMR vaccine) which is recommended as part of a child's immunization schedule. The use of the MMR vaccine is not without controversy; some research studies have linked the growing incidence of autism in children to the MMR vaccine. Other studies have concluded that there is no evidence, apart from coincidence, to link MMR with either autism or the bowel disorder Crohn's disease and parents are thus advised to have their children immunized with the triple vaccine.

The safest vaccines of all would be those that contain no DNA. A recent approach to vaccine development is to identify the specific protein antigen found in the pathogenic organisms and then to produce this antigen synthetically. Peptide vaccines based on the relatively short epitope sequences have limitations. Epitopes that elicit protective antibodies usually have conformations which depend on the particular folding of the polypeptide chains and it is difficult to reproduce these conformational epitopes in small synthetic peptides. Vaccines based on the intact protein antigen have, on the other hand, proved more effective and are extremely safe. There is no possibility of contaminating a synthetic protein vaccine with pathogenic organisms and moreover it is impossible to induce the disease state as the vaccine consists of a singe antigenic constituent and no nucleic acid material.

4.3.1   Hepatitis B vaccine

Hepatitis B is a liver infection caused by the hepatitis B virus. Liver cell damage in hepatitis B infection is due to an immune attack by T cells against

```
1                      25                          50
MENITSGFLGPLLVLQAGFFLLTRILTIPQSLDSWWTSLNFLGGSPVCLGQNSQSPTSNH
              75                         100
SPTSCPPICPGYRWMCLRRFIIFLFILLLCLIFLLVLLDYQGMLPVCPLIPGSTTTSTGP
      125                     150                  175
CKTCTTPAQGNSMFPSCCCTKPTDGNCTCIPIPSSWAFAKYLWEWASVRFSWLSLLVPFV
                  200                      226
QWFVGLSPTVWLSAIWMMWYWGPSLYSIVSPFIPLLPIFFCLWVYI
```

(4.3) HBsAg

viral antigens expressed on the surface of infected hepatocytes. The virus is a 42 nm spherical particle with a circular, largely double-stranded, DNA genome of approximately 3200 base-pairs at its core. Surrounding the nucleic acid is a protein shell and the surface of the virus consists of a protective phospholipid membrane studded with proteins (Figure 4.5). This protective envelope contains the surface antigenic determinant (HBsAg) (4.3), a 226 amino acid protein. Both glycosylated and non-glycosylated forms of the protein occur in the phospholipid envelope.

Infection with the hepatitis B virus leads not only to the production of the 42 nm particles but also to 22 nm particles that only contain the elements of the surface envelope. These latter particles are non-infectious because of the absence of nucleic acid and are the basis of the hepatitis B vaccine. Traditionally, the HBsAg particles were isolated from the plasma of infected individuals but a more recent source is to produce the particles in yeast cells by using recombinant DNA techniques. The HBsAg gene is encoded within a 889 base fragment of genome and the plasmid for expression contains the HBsAg coding sequence linked to the yeast alcohol dehydrogenase promoter.

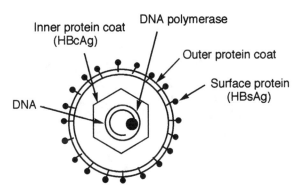

Figure 4.5 The hepatitis B virus structure. The viral particle is a 42 nm spherical double-shelled structure. The outer shell is a lipid envelope containing the hepatitis B surface antigen (HBsAg) and this surrounds an inner 27 nm icosohedral nucleocapsid composed of the hepatitis B core antigen (HBcAg). The nucleocapsid contains the genomic DNA (2200 kDa) and a DNA polymerase. The DNA is not totally double-stranded, but contains a large single-stranded gap of variable length. The DNA polymerase reaction appears to repair the gap

Expression in the yeast *Saccharomyces cerevisiae* leads to the production of protein which assembles into 22 nm particles similar to those made by human carrier patients. The recombinant particles are not identical to those isolated from plasma; they do not contain significant quantities of the glycoprotein implying that glycosylation is not required for immunity. The absence of intact hepatitis B virus and human proteins in the recombinant particles eliminates the possibility of secondary infections or autoimmunity problems as a result of the vaccine. There is no difference in the quantity, quality, or specificity of the antibody response induced by the recombinant vaccine compared with that derived from plasma.

The recombinant HBsAg vaccine was licensed for general clinical use in 1986, only a few years after identification, in 1979, of the surface antigen gene and its cloning and expression in yeast in 1982.

4.3.2 Approaches to an HIV vaccine

Human immunodeficiency virus (HIV) is the causative agent of acquired immune deficiency syndrome (AIDS). AIDS kills helper T cells (T_h), thereby crippling the immune system and rendering the patient susceptible to infection by microorganisms that rarely affect normal individuals. HIV (Figure 4.6) is an example of a retrovirus, i.e. the viral genome consists of RNA. Such viruses replicate by transcribing the RNA to double-stranded DNA (using reverse transcriptase, an enzyme carried by the virus), inserting the retrovirus genes into the host genome and then, by using the host's transcription and transla-tion machinery, this integrated DNA template produces the proteins and the

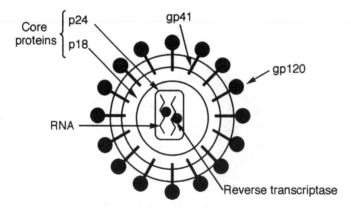

Figure 4.6   Structure of the HIV-1 virus. The virus particle is a sphere that is roughly 1000 Å across. The HIV envelope consists of two glycoproteins, gp120 and gp41, which are held together by non-covalent interactions. The membrane-and-protein envelope covers a core made of proteins p24 and p18. The viral RNA is carried in the core, along with several copies of the enzyme reverse transcriptase, which catalyses assembly of the viral DNA

RNA copies of the genetic material required for formation of new virus particles (Scheme 4.4). Once the T_h cells are infected, viral replication begins very quickly. The immune system responds to neutralize the infection but is eventually swamped and gradually becomes deficient due to the fall in the number of T_h cells. It is at this stage that the HIV positive individual becomes highly vulnerable to infection.

Unlike the body's reaction to most acute viral infections, the natural immune response does not destroy HIV. Many AIDS patients exhibit high levels of circulatory anti-HIV antibodies but these do not halt the progression of the disease. Thus, production of antibodies is not sufficient to combat infection successfully. Most viral vaccines activate the antibody arm of the immune system, so stimulating the formation of protective antibodies. No

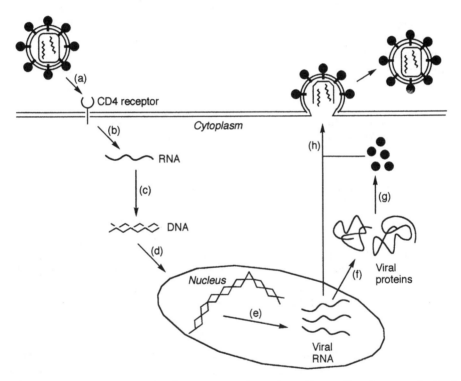

Scheme 4.4 Replication cycle of HIV. (a) Binding of gp120 to cell surface protein CD4 and fusion of the viral envelope to the host cell membrane. (b) Uncoating of the virion – the nucleocapsid complex and viral enzymes are released. (c) Reverse transcription of the viral RNA into a DNA copy and conversion of the newly synthesized DNA strand into double-stranded DNA. (d) Transportation of the viral DNA to the nucleus where it integrates at random sites in the host chromosome (pro-viral DNA). (e) Transcription of the pro-viral DNA to produce many copies of viral RNA. (f) Translation of RNA into viral proteins. (g) Post-translational processing of proteins, e.g. protease cleavage to generate viral structural proteins. (h) Assembly and budding of new virus particle. *Note:* the lipid membrane is derived from the infected cell

vaccines have yet been designed to stimulate the other arm of the immune system – the cellular component. In cellular immunity, activated cytotoxic T cells recognize and eliminate virus-infected cells rather than the infective agent itself. The most effective HIV vaccines may be those that stimulate both antibody and cellular arms of the immune response generating antibodies and activated cytotoxic T cells. It may not be possible to completely prevent initial infection but a vaccine may be able to prime the immune system to attack HIV as soon as it appears, thus greatly reducing the level of free virus in the blood stream. If the virus can be kept at a low concentration, an infected carrier would be less contagious and the disease may never progress to AIDS. Vaccine-induced immunity may therefore succeed in containing the virus where the naturally infected body does not.

The standard type of viral vaccines for warding off disease may be inappropriate for HIV as the mechanism by which the virus destroys the immune system is poorly understood and this makes it difficult to predict what constitutes a 'safe' vaccine. A vaccine based on a wild-type virus is clearly inappropriate as recipients would almost certainly become infected. An attenuated live strain of HIV is an unlikely vaccine strategy as there is the possibility of the weakened virus reverting to the virulent form. However, by systematically deleting genes critical for HIV replication it may be possible to develop a variant of the virus that can elicit a strong immune response without giving rise to AIDS. Making a 'killed' virus vaccine is also a possibility but the process of inactivation may cause the virus to shed its envelope proteins which are a major target for protective immunity.

The envelope proteins are the virus' means of gaining entry into host cells and they are therefore targets for vaccines that stimulate the production of neutralizing antibodies. The glycoproteins of the outer virus envelope have two components, namely gp41 which spans the membrane and gp120 which extends beyond it. Gp120 is anchored to the virus by non-covalent interactions with gp41 and both are derived from viral glycoprotein gp160 by proteolytic cleavage of the latter during viral replication. The first step in the infection of a cell by HIV is the binding of gp120 to CD4 receptors on the surface of host T cells. Antibodies that prevent attachment of gp120 to CD4 should be able to prevent viral infectivity. The CD4 binding site on gp120 is an attractive candidate for vaccine development and both gp120 and gp160 have been tested as HIV vaccine candidates. Gp120 and gp160 proteins were found to successfully induce the production of antibodies that recognized the envelope proteins and prevent laboratory-grown HIV from infecting cultured cells. However, the antibodies were ineffective at neutralizing HIV strains isolated directly from infected patients. The tertiary structure of the envelope protein in laboratory-grown virus strains is looser than that of surface protein in patient isolates and the CD4 binding determinant is more efficiently presented on the virus or on virus-infected cells than on purified gp120. Antibodies to laboratory-grown strains of virus would therefore not 'see' their targets on HIV isolated from patients. Neither may the pure protein be a better way to stimulate antibody production; gp120 does not have a precise conformation and gp160 clumps into ineffective aggregates.

When the envelope protein binds to a cell, the gp120 changes shape. A vaccine that duplicates the conformation adopted by the protein as it attaches to receptors, rather than one which duplicates the conformation present on the virion surface, may succeed better at raising antibodies that are able to block HIV from infecting human cells. Envelope proteins could be presented in a more natural conformation to the immune system by embedding them in 'pseudovirions' – artificial structures that resemble virus particles. The artificial viruses would be safer as vaccines than whole killed virus as they lack the genes that could propagate HIV infection, although they are difficult to produce in a stable form.

Another factor which may influence the design of a vaccine is the high mutation rate of the HIV virus; different HIV strains have envelope proteins which differ in amino acid sequences and tertiary structures. Thus, a vaccine that elicits production of antibodies against one strain of virus may not induce immunological protection against subsequent genetic variants.

Viral surface proteins do not generate activated cytotoxic T cells and they are therefore poor stimulants of cellular immunity. Cellular immunity is elicited through the binding of T cells to peptide fragments of the viral proteins that appear on the surface of an infected cell (see the following Section 4.4 for a detailed discussion of the cell-mediated immune response). For an HIV vaccine to stimulate cell-based immunity, the short pieces of foreign proteins would have to be displayed on the surface of cells. These cells would trick the body into mounting an immune response against all cells displaying the viral peptides, including those actually invaded by HIV. Inducing cells to produce and display HIV proteins may be achieved by engineering a benign non-HIV virus to carry genes encoding HIV-proteins. The carrier virus delivers the antigen-encoded gene to the host, where the antigen is then made. With *in situ* generation of the protein in the host it may be possible to induce cytotoxic T cell activity, thereby priming the immune response so that any cells that actually became infected with HIV would be killed. The most extensively tested live vector vaccines to date are based on the canarypox virus, a non-pathogenic relative of the smallpox virus. Genes that direct the production of gp160, gp120 and a variety of non-surface HIV proteins such as Gag, the core protein and protease, have been inserted into the canarypox virus. Live vector viruses tested to date in humans have elicited modest T cell-based immune responses and moreover have proved to be safe.

The genetic variability of HIV may preclude the development of a practical vaccine for AIDS. If vaccination against HIV does prove to be possible, the most effective vaccines are likely to be a prime boost combination to stimulate both arms of the immune response. For example, a patient might receive the canarypox vector–gp120 combination: the live virus vector to prime the immune system by generating cytotoxic T cells, and the protein to boost the response by eliciting antibody production.

4.4   The Cell-Mediated Immune Response

The cell-mediated immune response is the second branch of the immune system and involves specialized white blood cells known as T lymphocytes. The T cells are produced in the thymus and migrate to secondary lymphoid organs where they react with foreign antigen on the surface of other host cells. Some T cells kill virus-infected cells but the majority of T cells play a regulatory role in immunity by helping B cells to make antibodies. The diverse immunological responses are mediated by different classes of T cells:

(1) *Cytotoxic T cells (T$_c$)* These are lymphocytes which directly destroy cells which have been infected with viruses. The latter proliferate inside cells and cannot be attacked by antibodies. Cytotoxic T cells kill the infected cell before virus assembly can take place. The target cell is killed by making the membrane 'leaky' through formation of trans-membrane channels or by activating a self-destruct mechanism which causes apoptosis.

(2) *Helper T cells (T$_h$)* These are the 'master switches' of the immune system with a crucial regulatory role. T$_h$ cells secrete a variety of soluble mediators known as cytokines that help B cells make antibody responses. Cytokines also stimulate activated T cells to proliferate and activate natural killer (NK) cells and macrophages, the professional digesters of foreign materials in the body.

(3) *Suppressor T cells (T$_s$)* These are lymphocytes which inhibit the responses to T$_h$.

   B cells and T$_c$ cells are both involved directly in defence against infection and are known as effector cells. T$_h$ and T$_s$ are known as regulatory cells as they control the activity of the effector cells. While B cells secrete antibodies that can act far away, both effector and regulatory T cells act at short range by interacting directly with the cells they kill or regulate. Consequently, T cells bind foreign antigen only when it is on the surface of another cell in the body. T cells recognize antigen when it is in association with a family of cell-surface glycoproteins known as the major histocompatibility complex (MHC).
   The MHC is at the hub of T-cell specificity. Cytotoxic and helper T cells do not recognize the same MHC molecules on the surface of cells and this difference reflects the differing functions of the two types of cell. There are two classes of MHC molecules (Figure 4.7). Both MHC glycoproteins are hetero-dimers with one antigen-binding site; a groove formed from the variable domains and located at the end of the molecule furthest from the membrane. The MHC class I glycoprotein is expressed on almost all nucleated cells while class II has a more restricted tissue distribution and is confined largely to cells

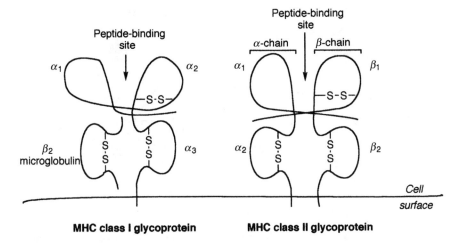

Figure 4.7   Structure of MHC molecules highlighting homologies. Both molecules are heterodimers with four extracellular domains each comprising approximately 90 amino acid residues. Three of the domains have intra-chain disulphide bonds. The two domains closest to the membrane are immunoglobulin-like and invariant, while the two domains furthest from the membrane are extremely polymorphic, i.e. variable, bind foreign antigen and present it to T cells

(macrophages and B lymphocytes) involved in immune responses. Class I molecules bind peptides from *intracellular* proteins such as viral proteins synthesized within host cells, while class II molecules bind peptides from *extracellular* proteins that become internalized. It is the complex of the MHC molecule plus foreign antigen which is recognized by the T cell. Those latter cells which recognize a cell expressing antigenic peptide in association with the MHC class I generally results in cytolysis of the cell presenting the antigen and thus these cells are cytotoxic T cells. As class I molecules are on most cells, this allows most cells to signal to cytotoxic T cells when they have been invaded. In contrast, T cells which recognize foreign antigen in association with MHC class II on the surface of specialized antigen-presenting cells (APCs) perform a regulatory role in the immune response to an antigen and are thus known as helper T cells (Figure 4.8). If the functional dichotomy between T cell recognition of peptides presented by the different MHC glycoproteins is not preserved then binding of peptides degraded from extracellular proteins would result in the killing of macrophages and B cells by cytotoxic T cells.

Binding of antigen–MHC on the target cell to receptors on the T cell is not strong enough to mediate a functional interaction between the two cells. Additional adhesion proteins or co-receptors are needed to stabilize the interaction by increasing the overall strength of cell–cell binding. The co-receptor on T_c is the CD8 glycoprotein, while the CD4 molecule provides the necessary stabilization on T_h. Both co-receptors have extracellular domains that bind to the invariant parts of the MHC molecule. Thus, CD8 T_c cells

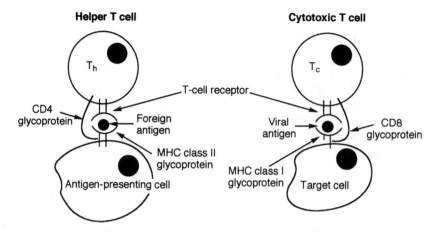

Figure 4.8   Interaction of T cells, MHC molecules and antigen. Cytotoxic T cells recognize foreign viral antigen in association with MHC class I molecules on the surface of any host cell, whereas helper T cells recognize foreign antigens in association with MHC class II molecules on the surface of an antigen-presenting cell

recognize MHC class I molecules and kill virus-infected cells, whereas CD4 T_h cells recognize MHC class II and selectively stimulate other types of cell to respond to antigen.

The effect of peptide binding to an MHC molecule is that its conformation is fixed to correctly expose the amino acid side-chains for recognition by the T-cell receptor (TCR). The latter is a membrane-bound antibody-like hetero-dimer (Figure 4.9). The receptor molecule is composed of two disulphide-linked polypeptide chains, α and β, each of which is approximately 280 amino acids long. As in antibodies, the chains have a variable (V) amino-terminal

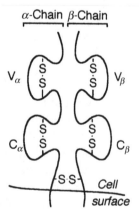

Figure 4.9   Structure of the T-cell receptor. The TCR resembles a membrane-bound Fab fragment and as in such fragments the juxtaposition of the variable regions forms the antigen-recognition site

region and a constant (C) carboxyl region but, unlike antibodies which have two antigen-binding sites for antigen, the TCR has only one, formed by the interaction of the V_α and V_β domains. The TCR is physically associated on the cell surface with an invariant set of polypeptide chains called the CD3 complex. The latter complex is thought to be involved in passing the signal from an antigen-activated TCR to the cell interior.

In contrast to antibodies which recognize antigenic determinants of a protein in their native conformation, T cells recognize epitopes on the *unfolded* polypeptide chain. The antigens seen by a T cell are degraded fragments of a foreign protein. The protein is degraded inside a virus-infected cell or an APC and then peptide fragments are transferred to the cell surface where they associate with MHC glycoproteins. The viral antigens presented to T_c cells are fragments of a protein synthesized within an infected cell, while antigens presented to T_h are peptides resulting from the processing of protein which has been ingested by the APC (Scheme 4.5). A T_c cell recognizes protein fragments complexed to class I MHC molecules and kills the infected cell. The binding of foreign peptides to MHC class II activates T_h cells. How activated T_h cells play a regulatory role in the immune response and help B cells make antibody responses is a complex process involving various secreted proteins called cytokines.

Activation begins when the T cell stimulates the APC to secrete interleukin-1 (IL-1). The combination of antigen binding plus IL-1 causes the T cell to secrete a growth factor called interleukin-2 (IL-2) as well as to synthesize cell surface IL-2 receptors (IL-2R). The binding of IL-2 to its receptors stimulates the T cell to proliferate (Scheme 4.6). Once activated a T_h cell can influence the function of other lymphocytes such as B cells. Activation of these cells by T_h cells requires that the cells be physically linked. This is achieved by both the B and T cells interacting with the same antigen, albeit at different epitopes. Thus, antigen is bound to specific antibodies on the surface of B cells and the antigen–antibody complex is then internalized. The antigen is degraded and recycled to the B-cell surface as peptides bound to MHC class II molecules.

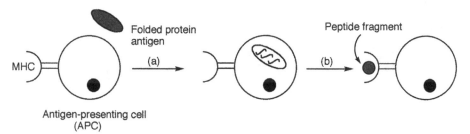

Scheme 4.5 Antigen processing. (a) Foreign protein ingested and partially degraded. (b) Some fragments expressed on the cell surface in association with MHC class II glycoprotein. Dendritic cells in particular are extremely efficient APCs because they express high levels of MHC class II molecules, co-stimulatory molecules such as B7 and other adhesion molecules

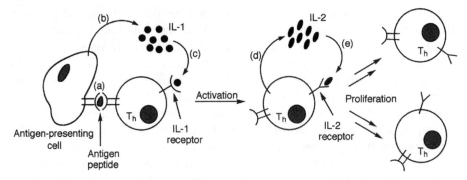

Scheme 4.6 T-cell activation and proliferation. T-cell activation results in proliferation which expands clones of cells, each of which has the same antigen specificity, to augment the immune response to the presented antigen. (a) Binding of T cell to antigen on surface of APC; (b) secretion of IL-1 by APC; (c) binding of IL-1 to its receptor; (d) secretion of interleukin IL-2; (e) binding of IL-2 to its receptor

Peptide–MHC class II complexes can interact with a T_h cell with the appropriate TCR and this causes the T cell to secrete interleukin 4 (IL-4) which binds to receptors on the B cell. Once the B cell is activated, other T_h cell-derived interleukins (IL-5, IL-6) induce the B cell to proliferate and mature into an antibody-secreting cell (Scheme 4.7).

Interfering with the T-cell mechanism provides an enormous opportunity for specific immunotherapy. By administering agents that selectively either augment or suppress the host immune response, various disease processes can be moderated.

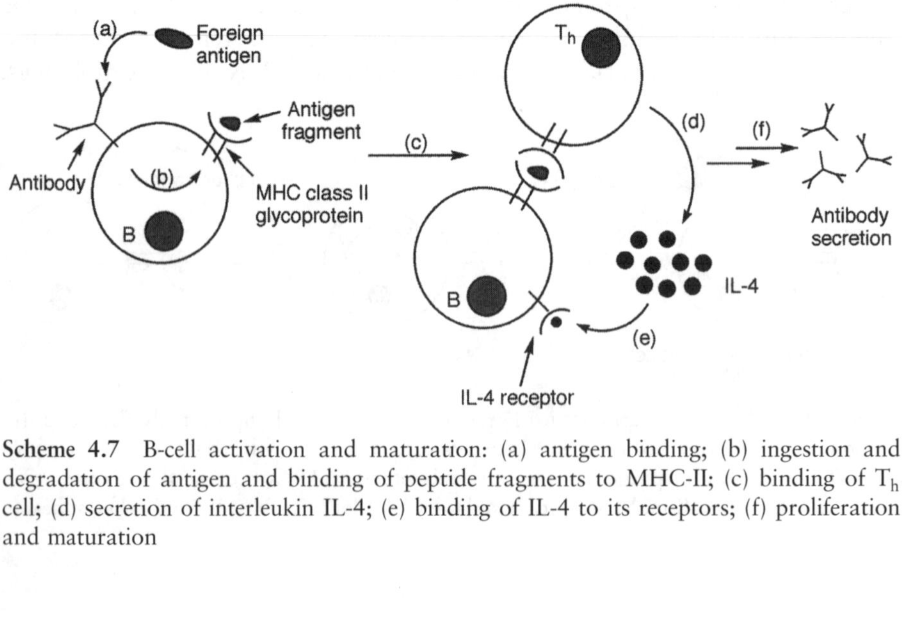

Scheme 4.7 B-cell activation and maturation: (a) antigen binding; (b) ingestion and degradation of antigen and binding of peptide fragments to MHC-II; (c) binding of T_h cell; (d) secretion of interleukin IL-4; (e) binding of IL-4 to its receptors; (f) proliferation and maturation

4.4.1   Immunoenhancement

The pivotal role of the interleukins, particularly IL-2, in the cellular immune response has made these molecules important therapeutic targets. IL-2 is a 133 amino acid globular protein with a molecular mass, depending on the degree of glycosylation, of 15–18 kDa and a structure composed of a compact core of four antiparallel α-helices connected by three loops. Although IL-2 is produced by antigen-stimulated T_h cells, it induces other lymphoid cells to proliferate and differentiate by binding to cell surface IL-2 receptors. These receptors are constitutively expressed by NK cells, lymphocyte like cells which kill some types of tumour cell and some virus-infected cells. The introduction of IL-2 in the absence of antigen in short-term cultures results in the selective stimulation and proliferation of NK cells while most T cells and B cells remain unchanged. Administration of IL-2 can therefore selectively and effectively boost the host immune system.

Immunostimulatory therapy with IL-2 has found use in the treatment of some cancer patients (see Section 4.5.2 below). T lymphocytes from a patient's tumour are harvested, cultured for several days in the presence of recombinant IL-2 to promote the proliferation of NK cells, and then at the end of this period the cells, which now contain billions of NK cells, are reinfused into the patient. There are side-effects (fever, chills, general malaise, erythema-rash, nausea-vomiting and hypertension) associated with the procedure resulting from high doses of IL-2 but, depending on the tumour type, partial or complete responses have been seen in 13 to 57% of patients with advanced cancer. IL-4 (a 20 kDa glycoprotein) has also been introduced as an immune system stimulator in various cancer treatment regimens.

4.4.2   Immunosuppression

Selective immunosuppression holds promise for the treatment of autoimmune diseases such as multiple sclerosis and rheumatoid arthritis and for improving the survival of organ transplants. Autoimmunity results from a breakdown in the mechanism that controls tolerance to self-antigens and thus in effect the immune system attacks itself. The most effective and selective immunointervention in autoimmune disease would be to establish or re-establish tolerance to self antigens. This requires knowledge of autoantigens which, at present, are poorly defined in most autoimmune diseases. In organ transplantation, the grafted tissue is usually recognized as foreign and rejected by the immune system of the recipient. Rejection is elicited by MHC molecules. The MHC genetic loci are extremely polymorphic and it is therefore quite unlikely that any two unrelated individuals will have an identical set of MHC glycoproteins. By suppressing the immune system, anti-donor responses can be dampened and thus allow the organ graft to be accepted by the recipient.

Combinations of several immunosuppresive drugs are normally used to prevent graft-vs-host disease in patients who have received allogenic transplants; this reduces the direct toxicity of individual drugs while maintaining the same overall efficacy. Combination drug regimens tend not to be used in autoimmune disease because of the risk of overimmunosuppression.

There are several approaches to suppressing the cellular immune response. The common goal is to selectively interrupt the activation of CD4 T cells by, for example, blocking the MHC–peptide–TCR interaction or interfering with the signal transduction pathway for IL-2R and IL-2 gene expression (see below). Alternatively, activation may be prevented by blocking the CD4 and other accessory molecules on the T cells or inhibiting binding of IL-2 to its receptor (Figure 4.10). Given the large number of surface molecules involved in CD4 T-cell activation, mAbs are attractive molecules with which to interfere with the activation process.

The anti-CD3 monoclonal antibody muromonab-CD3, a murine immunoglobulin produced by the hybridoma technique (see Section 4.2.2 above), recognizes, binds and blocks the CD3 complex of the TCR. This murine mAb was the first to become available for therapy in humans and is a component of triple and quadruple drug regimes used to prevent tissue rejection in kidney transplants. Muromonab has a number of side-effects, including overimmunosuppression and the development of human antimurine antibodies that lead over time to neutralization of the antibody.

Targeting more restricted T cell markers such as IL-2R, which is expressed only on activated but not resting T cells, generates more specific and better tolerated immunosuppressive agents. The IL-2R consists of three chains (α, β and γ). The α-chain is expressed only after T-cell activation and mAbs directed against the α-chain can specifically block the receptor. The first generation of anti-IL-2R monoclonal antibodies consisted of mouse and rat antibodies.

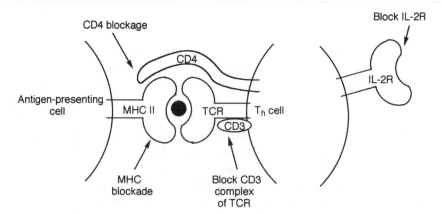

Figure 4.10   Targets for selective immunosuppression: TCR, T-cell receptor; CD, cluster of differentiation; MHC II, major histocompatability complex, class II; IL-2R, interleukin 2 receptor

These were promising but their immunogencity, short half-life and inability to recruit host effector functions limited their clinical use. Chimerization or humanizing these antibodies resulted in antibodies with a predominantly human framework which retained the antigen specificity of the original rodent monoclonal antibody. The fully humanized anti-IL-2R mAb, daclizumab, and the chimeric anti-IL-2R mAb, basiliximab, are well tolerated and effective in the immunoprophylaxis of patients undergoing renal transplantation.

Co-stimulatory and adhesion molecules are also important targets for intervention. In addition to CD4, which is the co-receptor for the MHC II glycoprotein, other adhesion molecules on the T cell include CD28 and leukocyte function-associated antigen 1 (LFA-1) which interact with the APC ligands B7 and intracellular adhesion molecule 1 (ICAM-1), respectively. Clinical trials with, for example, a murine $F(ab')_2$ anti-LFA-1 antibody are currently in progress for the suppression of rejection following organ transplantation. Monoclonal anti-CD4 antibodies have been used in clinical trials with a limited series of patients with autoimmune diseases such as rheumatoid arthritis and multiple sclerosis but it is premature to interpret the data. Multiple sclerosis is a chronic inflammatory debilitating disease of the nervous system that is thought to have both a genetic and environmental aetiological background. Relapsing–remitting multiple sclerosis, in which a patient experiences increasingly frequent relapses and intervening periods of remission, is one of several forms of the disease. Damage to the myelin sheath and other components of the nervous system is thought to be implicated in the pathogenesis of the disease; myelin autoantigens may activate the cytokine network resulting in excessive immune responses. The frequency and severity of disease relapses in patients can be decreased by β-interferon, one of a family of multi-functional proteins which *inter alia* mediate anti-proliferative activities.

Ligation of the TCR to the peptide-MHC complex is the crucial stimulus which triggers the series of intracellular events culminating in the transcription of genes which encode for IL-2 and its receptor. Blocking transmission of the signal between the molecules at the cell surface and the nucleus offers another way of manipulating T-cell-mediated immune responses. Natural product screening programmes led to the identification of the undecapeptide cyclosporin A (CsA) (**4.4**) and the macrocyclic lactone tacrolimus (FK506) (**4.5**) as compounds which can inhibit the production of IL-2 at the transcriptional level (Scheme 4.8). Rapamycin (**4.6**), another macrocyclic lactone, has no effect on the production of IL-2 but it potently inhibits a later lymphokine-associated signalling pathway.

Compounds **4.4** to **4.6** are inactive themselves but form pharmacologically active complexes with members of a family of intracellular receptors, the immunophilins. The cyclophilin family of immunophilins binds CsA, whereas the FK506-binding protein (FKBP) family binds the structurally related tacrolimus and rapamycin. The CsA•cyclophilin and FKBP•tacrolimus

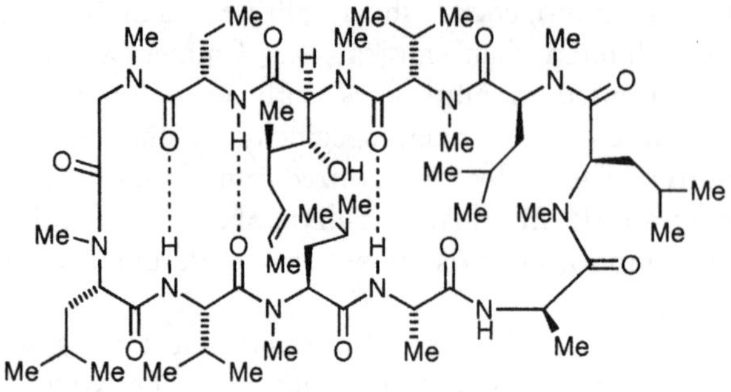

(4.4) Cyclosporin A

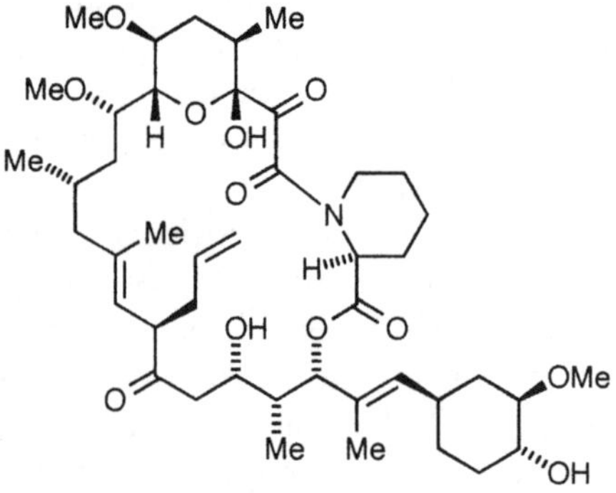

(4.5) Tacrolimus (FK506)

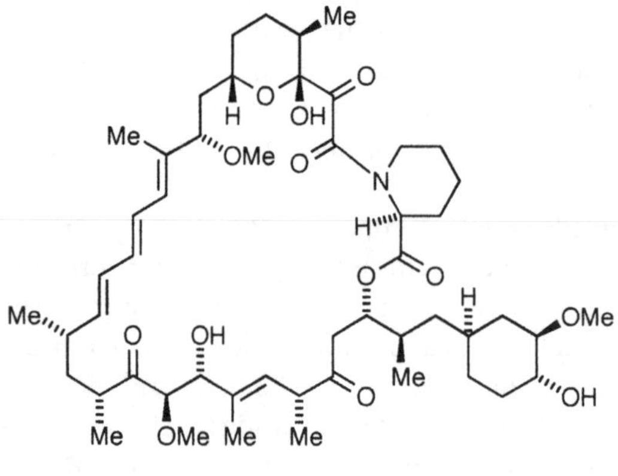

(4.6) Rapamycin

complexes specifically bind to, and inhibit, the enzyme calcineurin, a calcium-dependent phosphatase, which is required for transcriptional activation of the IL-2 gene. The inhibition of IL-2 transcription by tacrolimus and CsA is the result of indirect inhibition of the dephosphorylation of the calcineurin substrate NF-AT, a nuclear transcription factor. Although tacrolimus and rapamycin both bind to FKBP, rapamycin does not inhibit the same TCR-

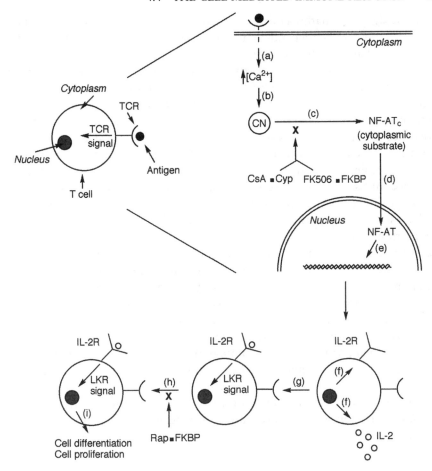

Scheme 4.8 Early events of the T cell activation cascade and the sites of inhibitory action by CsA, tacrolimus and rapamycin. (a) Stimulation of the T-cell receptor (TCR) by foreign antigen presented by an MHC molecule on the surface of an APC. (b) This leads to a rise in the concentration of intracellular Ca^{2+} which activates calcineurin (CN). (c) Calcineurin dephosphorylates the nuclear transcription factor of activated T cells (NF-AT) thus allowing (d) its translocation to the nucleus. NF-AT is one of the many transcriptional factors required for antigen-induced gene expression. (e) Binding of nuclear NF-AT to the enhancer of the IL-2 gene resulting in transcription. (f) Translation of the resultant message is followed by secretion of IL-2 and expression of IL-2R on the surface of the cell. (g) After the binding of IL-2 to the IL-2R, receptor a lymphokine receptor (LKR) signal transmission pathway (h) is activated. (i) Transduction of this signal through the cytoplasm and into the nucleus where a different set of genes is transcribed and translated.

CsA and FK506 (tacrolimus) inhibit the TCR-mediated signal transduction pathway through their ability to modulate the phosphatase activity of calcineurin. The calcineurin–CsA•cyclophilin and calcineurin–FK506•FKBP complexes interfere with the dephosphorylation of NF-AT, thus preventing its nuclear translocation. Rapamycin acts at a later stage of T cell activation by inhibiting Ca^{2+}-independent signalling pathways that emanate from the lymphokine receptor IL-2R. The target of the rapamycin•FKBP complex is not known but it is speculated to be a kinase or phosphatase (FKBP, FK506 (tacrolimus) binding protein) mediated signalling pathway that is affected by tacrolimus and CsA. Rather, it blocks a later Ca^{2+}-independent pathway associated with T-cell activation, which is mediated by the IL-2 receptor.

CsA was introduced in the early 1980s and has shown notable efficacy when combined with corticosteroids as a primary immunosuppressive therapy in treatment of patients after liver and kidney transplants. Tacrolimus has been introduced more recently and has more potent immunosuppressive properties compared with cyclosporin A. Both drugs are toxic to kidney and nerve cells, effects which are reversible in most instances. The introduction of tacrolimus-based therapy has seen a significant reduction in the incidence and severity of rejections of heart, lung and liver transplants, and moreover the need for concomitant use of corticosteroids is reduced.

The use of cyclosporin A, tacrolimus and the anti-lymphocyte antibody preparations have been significant factors in the gradual improvement of organ graft survival rates. Transplant patients receiving immunosuppressant therapy are however at increased risk of developing cancer, particularly lymphomas. New drug combinations and immunosuppressive strategies continue to evolve in order to achieve more effective control of rejection while minimizing injury to the graft and risk to the patient. While effective as immunosuppressants for the treatment of life-or-death situations such as organ transplantation, the shortcomings of CsA and tacrolimus make them less suitable for non-acutely life-threatening conditions such as autoimmune diseases.

As the immune system is so central to human pathologies, treatment will continue to be developed to programme it (vaccines), stimulate it (cytokines) and inhibit it.

4.5   Cancer Immunotherapy

Cancer is not a single disease but many, and therein lies the problem. There are approximately 200 types of cancer and each requires a different method of treatment. Cancer arises from cells in the body that were once normal cells but which have grown in an uncontrolled manner forming lumps or tumours. Abnormal cell proliferation is caused by defects in the genes that accelerate cell growth (oncogenes) and that slow down cellular turnover (tumour suppressor genes). The mutated genes which underlie the disease are a result of multiple events caused by viruses, chemicals or radiation, with probably four to six genetic changes being necessary to produce human cancers.

Many tumours occur with great frequency but pose no problems as they are benign and localized, e.g. warts. The problems start when the transformed malignant cells metastasize i.e. spread to other parts of the body and set up secondary areas of invasive growth. As cancer is not a single disease, there is not a single treatment for all cancers. The main 'curative' therapies are surgery and radiation but these are generally only successful if the cancer is found at an early stage. Once the disease has progressed to locally advanced cancer or metastatic cancer, these therapies are less successful. For many cancers some form of systemic therapy is required. Cancerous cells differ only subtly from cells in normal tissue but it is these differences which can be exploited in the development of anticancer agents. Chemotherapy is a mainstay of treatment in most solid tumours, e.g. carcinoma of lung, colorectal and advanced breast cancer. One of the hallmarks of malignant tumours is their high cell proliferation rate and many cytotoxic drugs, e.g. bleomycin, etoposide, cisplatin, cyclophosphamide, doxorubicin, 5-fluorouracil and methotrexate, modify DNA which impairs accurate replication.

The genetic changes which transform healthy cells into cancerous ones cause alterations in cell surface biochemistry. Cancer cells may, for example, display altered cell surface glycoproteins and glycolipids that may be recognized by the body as foreign and/or produce tumour-associated antigens (TAAs) which identify cells as abnormal and supposedly pave the way for their destruction by phagocytosis or some branch of the immune system (Scheme 4.9). All humans produce numerous abnormal cells each day but most individuals do not suffer from cancer. Somehow, in cancer, the body's immune system fails to destroy unwanted cells and thus a tumour results. An ineffective immune response may be a consequence of tumour cells displaying a low epitope density at the cell surface which prevents immunocompetent cells from recognizing the cell as unusual and destroying it. Antibodies and cytotoxic cells may also have difficulties accessing the surface of tumour cells within a mass. Alternatively, the epitope structures which identify the abnormal cell may be camouflaged so the immune system cannot recognize the cells. Another possibility for a failed immune response is that cancerous cells may produce a substance which may induce some degree of immune suppression.

Although most tumour cells are poorly immunogenic, this does not mean that they lack potential antigens that could be used to induce an anti-tumour immune response. The goal of cancer immunotherapy is to augment the body's own immune system and enhance an intrinsic weak specific antitumour response.

4.5.1 Antibody-mediated immunotherapy

Many tumour-associated antigens (TAAs) are the products of mutated oncogenes or tumour suppressor genes and are not normally present intact on the surface of the tumour cell. Extracellular antibodies would not be able to discriminate between cells bearing mutant or normal protein and are not likely to be of therapeutic value. In some cancers where antigens are expressed at higher density on malignant cells relative to normal cells, antibodies have been shown to be useful therapeutic agents. In 25–30% of breast and ovarian

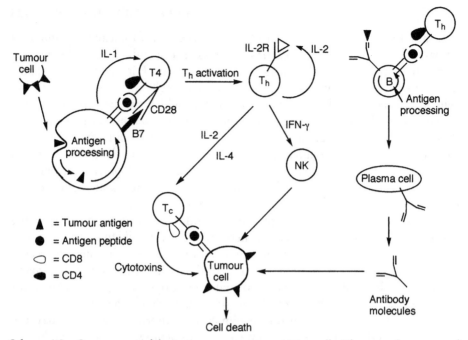

Scheme 4.9 Components of the immune response to tumour cells. There are three types of immune cells, i.e. B, T_c and T_h, that recognize tumour antigens and mediate tumour cell destruction. B cells and T cells see different epitopes on the same molecules. B cells recognize primarily unprocessed antigens in their tertiary configurations, although they frequently require the help of T cells for full antibody production. T cells recognize processed peptide presented in linear fashion in the MHC-binding groove. A tumour can present its antigens in two ways to T cells, i.e. (i) directly T_c cells as 8–10 amino acid chains in association with MHC class I molecules, or (ii) indirectly to T_h cells as 12–18 amino acid chains in association with MHC class II molecules after the antigens, released by cells destroyed by T_c cells, have been taken up by 'professional APCs'. Antigens presented to class I MHC are normally self-antigens, mutated self-antigens, or antigens of viruses or other intracellular pathogens. Most cells in the body express class I MHC and so are potential APCs. Tumour cells sometimes lack the MHC class I molecules responsible for presenting the relevant tumour antigen and so escape immune detection by T_c cells. Antigens presented to class II MHC molecules are extrinsic antigens such as those from most bacterial infections and standard vaccines. Class II molecules are normally present on professional APCs such as B cells, dendritic cells and macrophages. One of the critical features that defines a professional APC is the expression of membrane-bound co-stimulatory signals such as the B7 molecule, which binds to the CD28 receptor expressed by CD4 T_h cells. Occupancy of T-cell receptors in the absence of these co-stimulatory signals can lead to immunological tolerance. One reason why tumours are only weakly immunogenic is that although they may express TAA that can be recognized by T cells, they lack the co-stimulatory molecules which are required for full effector function. Binding of antigen-MHC-II complex to resting T_h (T4) cells and the release of IL-1 from the APC activate T_h cells. Such activated cells release various cytokines including interleukins 2 and 4 which stimulate proliferation and differentiation of B cells into antibody-producing plasma cells and which activate T_c cells for direct tumour lysis. Killing is caused by cytotoxins released by the activated T_c cells. Activated T_h cells also produce γ-interferon (IFN-γ) which fosters activation of NK cells, antigen-non-specific cytotoxic lymphocytes. A second antibody-independent mechanism of cytotoxicity involves the interaction between NK cells and tumour epitopes. The relative importance of T_h cells and T_c cells in host immunity against cancer appears to vary with different tumour antigens

cancers, the human epidermal growth factor receptor oncogene (HER2/neu, also known as c-erbB-2) is overexpressed and amplification of the HER2/neu protein is associated with poor clinical outcome. As the proliferation of cancer cells is regulated by growth factors and their receptors, oncoproteins such as HER2/neu are ideal targets to exploit in the search for new antitumour therapies. The humanized monoclonal antibody trastuzumab (herceptin) directed to an extracellular domain of the HER2/neu protein was licensed for the treatment of breast cancer patients in 1998. The antibody exerts its antitumour activity not only by blocking the receptor but also by effecting antibody cell-mediated cytotoxicity. Trastuzumab is active as a single agent in patients whose tumours overexpress HER2/neu but results are more impressive when used in combination with chemotherapy.

Haematological malignancies are ideally suited to mAb-based therapy because of the ready accessibility of neoplastic cells in the blood, bone marrow, spleen and lymph nodes that allow rapid and efficient targeting by specific mAbs. Antibodies work best when the targets are highly expressed. The B-cell marker CD20 is an example of such a molecule and impressive clinical results have been obtained with anti-CD20 (rituximab) therapy for B-cell lymphoma. Rituximab, a chimeric monoclonal antibody, specifically binds to the CD20 antigen on normal and malignant B cells. It produces antibody-dependent cellular- and complement-mediated cytotoxicity in these cells and in 1998 was licensed for the treatment of patients with low-grade B-cell non-Hodgkin's lymphoma.

Both trastuzumab and rituximab contain human Fc regions and can therefore recruit complement and cell effector functions. Despite being effective and well tolerated, 50% or more of patients do not respond to these antibodies. Neither of the two cytotoxic mechanisms by which antibodies mediate the elimination of cells expressing a target antigen, complement-mediated cytotoxicity and antibody-dependent cell-mediated cytotoxicity, are particularly efficient and for most tumour antigens binding by antibody is not sufficient to induce cell lysis. One approach to improve the efficacy of mAb-based immunotherapy is to link cytotoxic agents (toxins, anti-cancer drugs and radionuclides) to tumour-targeting mAbs (see Section 4.2.2 above). Radiolabelled antibodies address the problem of poor tumour penetration since the isotopes used emit β-radiation with a range of several centimetres. Thus, tumours may be exposed to a cytotoxic dose without requiring the antibody to penetrate to the tumour core. The mouse mAb parent of rituximab can be radiolabelled with ^{90}Y using 1-benzyl isothiocyanate-4-methyl-diethylenetriamine pentaacetic acid (4.7) and is specific for the CD20 antigen. Clinical trials of the ^{90}Y-anti-CD20 murine mAb in patients with recurrent B-cell non-Hodgkin's lymphoma showed a response rate of 72% (33% complete response) following a single infusion of the antibody. There is also evidence from a randomized trial that suggests that the radioimmunotherapy approach is superior to the use of 'naked' rituximab.

Chimeric toxins containing a growth factor instead of an mAb are often considered a type of immunotoxin. The recombinant protein DAB$_{389}$IL2

(4.7)

which contains the initiator methionine and the first 388 amino acids of diphtheria toxin fused to IL-2 has become a drug (denileukin difitox) that oncologists can prescribe for cutaneous T-cell lymphoma (see Section 3.4.2 earlier).

Another approach to circumvent the inefficient nature of antibody-mediated effector functions is to use bispecific antibodies (BsAbs) as a means to recruit the cellular arm of the immune response. In the prototypical BsAb, each half of the antibody structure is derived from a different monoclonal parent. Thus, one arm of the antibody Fv could be directed to a tumour antigen and the other to an effector antigen or molecule such as the T-cell receptor CD3 complex or the Fc receptor FcγRIII (CD16) on NK cells. These BsAbs are hybrid proteins which are constructed by the fusion of two hybridoma cell lines. Many other BsAb formats are possible, ranging from simple $F(ab')_2$ dimers by chemically cross-linking Fab fragments of two parent antibodies, to dimeric scFvs and tetravalent Ab-scFv molecules by elegant genetic engineering techniques. A few BsAbs have been tested in early clinical trials. Although evidence of biological activity was observed *in vivo*, the trials were limited by toxic effects. The efficacy of BsAb-based cancer treatments will probably ultimately depend on finding the right combination of trigger molecule and target antigen.

4.5.2   T-cell-mediated immunotherapy

As T cells play a key role in orchestrating the host immune system against cancer, an increased understanding of what is required to generate the T-cell response has opened up the possibility of using T-cell-based immunotherapy as a treatment for certain cancers.

Cytokine therapy

A defect in the helper arm of the immune response would result in failure to produce appropriate cytokines as a second signal to activate T_c cells. Antibody-induction is also highly dependent on T-cell help. Introducing a non-specific activator such as IL-2 into the patient would bypass the need for T_h cells and the release of cytokines. IL-2 is one of the most powerful activators of the immune system currently used in immunotherapy. Intravenous administration of IL-2 results in severe side-effects and the protein is toxic in high doses. IL-2 complexed to high-molecular-weight molecules such as poly(ethy-lene glycol) (see also Section 3.3.2 earlier) may circumvent many of the problems associated with the need for frequent parenteral administration. PEG-IL-2 is more soluble and has a much longer plasma half-life than IL-2 itself (947 min compared with 70–85 min for the parent drug), thus enabling its administration bi-weekly with results similar to those with twice-daily native IL-2 administration.

Effective results can also be achieved by transferring tumour-infiltrating lymphocytes (TILs) into patients. TILs are white blood cells that invade and kill early tumour cells in cancer patients but as the tumours spread the TILs become overwhelmed and are unable to eliminate the malignancy. TILs are removed from a patient's tumour, grown to large numbers in tissue culture in the presence of IL-2, and the expanded population of TILs is reinfused back into the patient together with low doses of IL-2. By passively providing activated T cells to the patient, this technique is known as adoptive immu-notherapy. In clinical trials, this approach has resulted in partial responses in patients with metastatic melanoma, colon and renal cell cancer.

Genetic modification of tumour cells can produce more defined biological effects. A missing function such as IL-2 may be replaced by inserting the cytokine gene into a patient's own tumour cells *ex vivo* so that when the cells are reintroduced into the patient the protein is produced at very high concentrations local to the tumour and the systemic concentrations are quite low. Calcium phosphate transfection is a well-established laboratory method for introducing foreign genes into cells; the calcium phosphate interacts with DNA to form complexes which are internalized by endocytosis. Tumour cells transfected with genes coding for cytokines, histocompatibility and co-stimulatory molecules have resulted in improved systemic antitumour immune responses in animal models. Most human tumours are difficult to transfect and the procedure is moreover highly individualized and impractical for treating large numbers of patients. Viral methods are a better way of getting genes into tumour lymphocytes from patients. Section 5.2 below on gene therapy discusses the technique of engineering cells to express genes encoding specific proteins in detail, but basically this entails altering a non-pathogenic virus to contain the gene of interest and then by means of the virus' normal infection machinery the therapeutic gene is carried into cells. A number of clinical trials testing the efficacy of genetically modified whole-cell tumour vaccines with genes that are known to be critical mediators of immune system activation (B7 and cytokine genes) are currently underway.

Vaccination

The idea behind cancer vaccines is that a tumour-specific preparation would prime T cells and thereby serve as an inducer for tumour-destructive immune responses. Most tumour antigens recognized by T cells are not known and early generations of cancer vaccines consisted of killed autologous tumour cells, i.e. the patients' own cells, together with an adjuvant, a compound that increases the immunogenicity of a given vaccine. Limited clinical success with this technique has been occasionally reported with melanoma and renal cancer patients. If, however, the TAA can be identified then vaccines comprising defined antigens could be designed. It may become possible to vaccinate patients whose tumours express a given antigen with the antigenic peptides or with the proteins contaning these peptides. Table 4.2 lists some candidate tumour antigens.

Many TAAs arise from mutated or rearranged oncogenes or tumour suppressor genes but the site of mutation or rearrangement tends to vary among tumours and therefore specific epitopes recognized by the immune system will differ from individual to individual which thus makes it difficult to produce a generic cancer vaccine. Mutations in the *p53* tumour suppressor gene have been implicated in a wide variety of human cancers, including leukaemias and lymphomas and carcinomas of the liver, bladder, brain, breast, lung and colon. The normal p53 protein is a transcription factor and is involved in cell cycle inhibition after DNA damage. This function is lost by many of the mutant forms of p53 found in human cancers. Many of the mutations in the *p53* gene are mis-sense and result in a protein which, although inactive, is expressed at significantly higher levels in tumours than in normal tissues. Wild-type p53 protein is composed of 393 amino acid residues with a molecular weight of 53 kDa. There are a few 'hot-spots' for mutation (such as codons 175, 248 and 273) but these make up only a small fraction of the mutations observed. This complicates mutation-specific immunotargeting, because each patient is likely to have a different mutation. The idea of a wild-type peptide as a vaccine is gaining momentum. As the mutant p53 protein is overproduced in many cancers, this might allow sufficient discrimination between tumour and normal cells to be of clinical utility. The

Table 4.2   Potential sources of tumour antigens

Category of antigen	Example
Oncogene product	• Ras; position 12 mutation in approximately 10% of all tumours
	• HER2/neu; overexpressed in 25–30% of breast and ovarian tumours
Tumour suppressor gene product	• p53; mutations in greater than 50% of human tumours
Reactivated embryonic protein	• Mage family; expressed in approximately 50% of human melanomas and in approximately 25% of human breast tumours
Viral proteins	• Human papillomavirus E6 and E7 gene products (cervical cancer)
	• Epstein–Barr virus (Burkitt's lymphoma and nasopharyngeal carcinoma)

major practical advantage of targeting a wild-type epitope is that a single vaccine preparation could be used to target a wide variety of human tumours producing different mutated p53 proteins.

Approximately 25% of breast and ovarian tumours are characterized by high levels of a growth factor receptor resulting from overexpression of the *HER2/neu* oncogene. The monoclonal antibody trastuzumab (see Section 4.5.1 above) directed to an extracellular domain of the receptor has already been licensed for use in patients with metastatic breast cancer but the oncogenic protein is also a target for vaccine-induced T-cell immunity.

Some TAAs are encoded by normal nonmutated cellular genes that are highly expressed in tumour cells but not in normal cells. For example, the *mage* gene family are expressed in melanoma and Mage-1 and Mage-3 peptides are two promising candidates for vaccine-based treatment of some patients with skin cancer.

Immunization protocols have involved administering the peptide epitope emulsified in incomplete Freund's adjuvant. The process of endocytosis introduces the peptide or protein into the cytosol of an APC where it associates with MHC class I molecules and is then expressed at the cell surface. In addition, the peptide can bind directly to the small fraction of empty MHC-I molcules that are already expressed at the surface of the APC. Peptide vaccination has only been tried on a limited scale in patients with cancer. Trials of the Mage-3$_{161-169}$ epitope (Ile-Met-Pro-Lys-Ala-Gly-Leu-Leu-Ile) in incomplete Freund's adjuvant reported promising preliminary clinical results in patients with advanced melanoma. HER2/neu-specific vaccine consists of several peptides, 15–18 amino acids long, mixed with an adjuvant and this formulation was found to induce T-cell immunity to the HER2/neu protein in patients with breast and ovarian cancer. Immunological responses have also been observed in cancer patients immunized with custom p53-derived peptides corresponding to the mutant sequence in each of their tumours. Inoculation with peptides in adjuvant does not always protect and with certain peptides vaccination has led to immunological tolerance rather than protective immunity.

Alternatively, the antigen can be delivered via a recombinant virus (see above). This approach is at an early stage of experimental and clinical investigation but it involves introducing the antigen into the viral genome of a non-pathogenic virus, e.g. vaccinia, so that infected cells will display the antigen together with the host MHC. Preliminary results of clinical trials with recombinant viral vaccines are just now beginning to be evaluated. Viral vaccines can also be modified to enhance the processing and presentation of encoded antigens by incorporating genes for cytokines such as IL-2 or co-stimulatory molecules such as B7.

Some cancers have been shown to result from viral infections, for example, the human papillomavirus (HPV) has been implicated in cervical cancer, hepatitis B with the development of liver cancer and the Epstein–Barr virus with Burkitt's lymphoma and nasopharyngeal carcinomas. While most cancer vaccines are therapeutic and involve attempts to activate systemic immune responses after, rather than before, the development of a tumour, there is potential for prophylactic vaccination to combat those tumours which are clearly induced by viruses. The incidence of hepatomas was found to decrease when preventative hepatitis B immunization was introduced in Taiwan, ostensibly to prevent hepatic damage caused by the virus. A high percentage of cervical cancers express E6 and E7 antigens and these proteins are targets for vaccines aimed at controlling HPV-induced tumours. Several experimental vaccine strategies have been developed to enhance cell-mediated immunity against cervical carcinoma. A recombinant vaccinia virus vector expressing HPV E6 and E7 and a specific peptide of the E7 antigen mixed with adjuvant have been put into clinical practice. Small-scale clinical trials using these vaccine strategies have shown encouraging results in patients with late-stage cervical cancer who had failed to respond to conventional therapy.

The discovery and identification of TAAs has made generic recombinant and peptide-based vaccines a possible alternative to autologous cellular cancer vaccines. Peptide-based vaccines are safe and can be synthesized with high purity and reproducibility, while recombinant viruses encoding TAA allow precise delivery of the antigen. Antigen heterogeneity is an inherent feature of malignancy and immunization against a single tumour antigen is not likely to be as effective as immunization against multiple tumour antigens. The goal is therefore to construct polyvalent vaccines against human malignancies to accommodate the heterogeneous expression of TAA on tumour cells and generate a stronger T_h and T_c response. It is often difficult to evaluate the efficacy of new treatments as the patients entering clinical trials are often those who have failed to respond to every conventional therapy and are thus not the best candidates. The first persons entering trials therefore provide much information for little personal benefit. Cancer therapy in the future may include peptide- or gene-based delivery of TAA in conjunction with cytokines. Therapeutic cancer vaccines illustrates how the concept of vaccines has widened to encompass non-infectious diseases and highlights the increasing blur between the areas of vaccines and gene therapy (see Section 5.2 below).

Further Reading

Textbook and review articles

- L. Adorini, J.-C. Guéry, G. Rodriguez-Tarduchy and S. Trembleau, Selective Immunosuppression, *Trends Pharmacol. Sci.*, 1993, **14**, 178–181.
- B. Alberts, D. Bray, J. Lewis, M. Raff, K. Roberts and J. D. Watson, *Molecular Biology of the Cell*, 3rd Edn, Garland Publishing Inc., New York, 1994.

- D. Baltimore and C. Heilman, HIV Vaccines: Prospects and Challenges, *Sci. Am.*, 1998, **279** (July), 78–83.
- E. Benjamini, R. Coico and G. Sunshine, *Immunology: A Short Course*, 4th Edn, Wiley, New York, 2000.
- I. Berkower, The Promise and Pitfalls of Monoclonal Antibody Therapeutics, *Curr. Opin. Biotechnol.*, 1996, **7**, 622–628.
- J. R. Bertino (Ed.-in-chief), *Encylopedia of Cancer*, Academic Press, San Diego, CA, 1996.
- M. Clark, Antibody Humanization: A Case of the 'Emperor's New Clothes'? *Immunol. Today*, 2000, **21**, 397–402.
- C. Dean and H. Modjtahedi, Monoclonal Antibodies, in *Molecular Biomethods Handbook*, R. Rapley and J. M. Walker (Eds), Humana Press, Totowa, NJ, 1998, pp. 567–580.
- S. J. DeNardo, L. A. Kroger and G. L. DeNardo, A New Era for Radiolabeled Antibodies in Cancer, *Curr. Opin. Immunol.*, 1999, **11**, 563–569.
- M. J. Glennie and P. W. M. Johnson, Clinical Trials of Antibody Therapy, *Immunol. Today*, 2000, **21**, 403–410.
- G. Kaplan, Z. A. Cohn and K. A. Smith, Rational Immunotherapy with Interleukin 2, *Bio/Technology*, 1992, **10**, 157–162.
- J. Kunz and M. N. Hall, Cyclosporin A, FK506 and Rapamycin: More Than Just Immunosuppression, *Trends Biochem. Sci.*, 1993, **18**, 334–338.
- A. Lanzavecchia, Identifying Strategies for Immune Intervention, *Science*, 1993, **260**, 937–952.
- M. A. Liu, Vaccines in the 21st Century, *BMJ*, 1999, **319**, 1301; www.bmj.com/cgi/content/full/319/7220/1301.
- S. Maulik and S. D. Patel, Re-Engineering the Immune Response, in *Molecular Biotechnology: Therapeutic Applications and Strategies*, Wiley, New York, 1997, pp. 154–185.
- R. Offringa, S. H. van der Burg, F. Ossendorp, R. E. M. Toes and C. J. M. Melief, Design and Evaluation of Antigen-specific Vaccination Strategies Against Cancer, *Curr. Opin. Immunol.*, 2000, **12**, 576–582.
- I. Pastan and D. FitzGerald, Recombinant Toxins for Cancer Treatment, *Science*, 1991, **254**, 1173–1177.
- D. M. Pardoll, Cancer Vaccines, *Nat. Med. Vaccine Suppl.*, 1998, **4**, 525–534.
- M. K. Rosen and S. L. Schreiber, Natural Products as Probes of Cellular Function: Studies of Immunophilins, *Angew. Chem. Int. Ed. Engl.*, 1992, **31**, 384–400.
- A. M. Scott and S. Welt, Antibody-based Immunological Therapies, *Curr. Opin. Immunol.*, 1997, **9**, 717–722.
- A. B. van Spriel, H. H. van Ojik and J. G. J. van der Winkel, Immunotherapeutic Perspective for Bispecifc Antibodies, *Immunol. Today*, 2000, **21**, 391–397.
- F. K. Stevenson, Tumor Vaccines, *FASEB J.*, 1991, **5**, 2250–2257.

- R. W. Tindle, Human Papillomavirus Vaccines for Cervical Cancer, *Curr. Opin. Immunol.*, 1996, **8**, 643–650.
- P. A. Trail and A. B. Bianchi, Monoclonal Antibody Drug Conjugates in the Treatment of Cancer, *Curr. Opin. Immunol.*, 1999, **11**, 584–588.
- T. A. Waldmann, Monoclonal Antibodies in Diagnosis and Therapy, *Science*, 1991, **252**, 1657–1662.
- E. J. Wawrzynczak, *Antibody Therapy*, BIOS Scientific Publishers Ltd., Oxford, UK, 1995.
- G. Winter and W. J. Harris, Humanized Antibodies, *Trends Pharmacol. Sci.*, 1993, **14**, 139–143.

Research publications

- L. K. Borysiewicz, A. Fiander, M. Nimako, S. Man, G. W. G. Wilkinson, D. Westmoreland, A. S. Evans, M. Adams, S. N. Stacey, M. E. G. Boursnell, E. Rutherford, J. K. Hickling and S. C. Inglis, A Recombinant Vaccinia Virus Encoding Human Papillomavirus Types 16 and 18, E6 and E7 Proteins as Immunotherapy for Cervical Cancer, *Lancet*, 1996, **347**, 1523–1527.
- M. L. Disis, K. H. Grabstein, P. R. Sleath and M. A. Cheever, Generation of Immunity to the HER-2/*neu* Oncogenic Protein in Patients with Breast and Ovarian Cancer Using a Peptide-based Vaccine, *Clin. Cancer Res.*, 1999, **5**, 1289–1297.
- P. T. Jones, P. H. Dear, J. Foote, M. S. Neuberger and G. Winter, Replacing the Complementarity-determining Regions in a Human Antibody with those from a Mouse, *Nature (London)*, 1986, **321**, 522–525.
- G. Köhler and C. Milstein, Continuous Cultures of Fused Cells Secreting Antibody of Predefined Specificity, *Nature (London)*, 1975, **256**, 495–497.
- P. Valenzuela, P. Gray, M. Quiroga, J. Zaldivar, H. M. Goodman and W. J. Rutter, Nucleotide Sequence of the Gene Coding for the Major Protein of Hepatitis B Virus Surface Antigen, *Nature (London)*, 1979, **280**, 815–819.
- P. Valenzuela, A. Medina, W. J. Rutter, G. Ammerer and B. D. Hall, Synthesis and Assembly of Hepatitis B Virus Surface Antigen Particles in Yeast, *Nature (London)*, 1982, **298**, 347–350.

5

Oligonucleotides

5.1 Overview

Therapeutic peptides and proteins are used to treat human disease at the metabolic or biochemical consequences of the underlying basic defects rather than directly at the causative agents themselves. The use of therapeutic oligonucleotides involves intervention at the nucleic acid level and, as with peptides and proteins, treatment can be either long-term or for a more restricted period.

Overexpression or activation of some genes is associated with a variety of malignant tumours. In Burkitt's lymphoma, for example, activation of the c-*myc* proto-oncogene results in unrestrained growth of tumour cells via an increase in c-Myc protein levels. Conceptually, such disease-associated proteins may be able to be blocked at the nucleic acid level by using oligodeoxynucleotides with base sequences complementary to those in the c-*myc* mRNA (antisense therapy). Such antisense agents form RNA–DNA hybrids that could block the expression of the oncogene thus preventing the production of the disease-associated protein. Interfering at the nucleic acid level to alter the course of cancer, for example, could also take the form of delivering a gene to tumour cells (gene therapy) that would down-regulate growth-promoting products. A suppressor gene such as *p53* or a gene coding for a cytokine that activates immune attack on the cancer cell could, in principle, slow the progression of a tumour.

Gene and antisense therapies hold enormous promise as molecular-based medicines but their development is at a stage similar to that of synthetic peptide and protein pharmaceuticals in the 1980s. There are many technical obstacles to be overcome before these genetic approaches will make a noticeable impact in the treatment of disease.

5.2   Gene Therapy

5.2.1   Introduction

Some diseases are due to a defect in a specific gene which in turn results in a deficiency of a specific protein, e.g. lack of factor VIII leads to haemophilia and insufficient α_1-antitrypsin causes emphysema (Table 5.1). As discussed in Chapter 2, the symptoms of this type of disorder can be alleviated by the administration of the deficient protein to those afflicted. Intervention at the oligonucleotide level involves delivering the deficient gene into the patient by using the technique of gene therapy. Thus, gene therapy has the potential to restore gene function that has been lost, instead of merely treating the symptoms.

While monogenetic defects associated with well-characterized inborn errors of metabolism and other simple genetic disorders are obvious clinical targets for gene therapy, the transfer of genetic information into defective or damaged tissue has potential application in the treatment of acquired disorders such as cardiovascular disease, cancer and neurodegenerative diseases; these are major causes of ill-health and death in the West (Table 5.2). Many forms of cancer are associated with, and possibly caused by, inappropriate regulation of gene expression of oncogenes and tumour-suppressor genes. By delivering a functional tumour-suppressor gene to diseased cells, for example, it may be possible to stop or reverse the malignant process. Another possible cancer treatment using the gene therapy approach could involve the introduction of genes encoding cytokines to activate and focus the immune response on to the tumour. Genetic manipulation to restore normal regulation of a host immune system also offers a therapeutic avenue to the treatment of patients with compromised immune responses such as those with autoimmune diseases and AIDS.

Thus, the broad definition of gene therapy is insertion of a functional gene into the somatic cells of a patient to correct an inborn error of metabolism, or to repair an acquired genetic abnormality, or to provide a new function to a cell. Gene therapy is thus a form of drug delivery in which altered cells

Table 5.1   Examples of diseases caused by single-gene defects

Disease	Defective gene
Cystic fibrosis	Cystic fibrosis transmembrane conductance regulator (CFTR)
Emphysema	α_1-Antitrypsin
Haemophilia	Clotting factors VIII, IX
Gaucher's disease	Glucocerebrosidase
Severe combined immunodeficiency disease	Adenosine deaminase
Pituitary dwarfism	Growth hormone

Table 5.2   Principal acquired diseases amenable to gene therapy

Disease	Gene inserted
Cancer	Interleukins, tumour suppressor genes, HSV-TK[a]
Cardiovascular diseases	
• Myocardial infarcts	tPA
• Prevention of blood clots	tPA
Neurodegenerative diseases	
• Alzheimer's disease	Nerve growth factor
• Parkinson's disease	Tyrosine hydroxylase
Autoimmune disease	
• Rheumatoid arthritis	Cytokine (IL-1) antagonists
AIDS	HIV antigens, cytokines, TK genes

[a]HSV-TK, herpes simplex virus thymidine kinase, an enzyme that makes cells much more susceptible to the antiviral drug ganciclovir.

produce the needed proteins continuously at therapeutic levels with concomitant significant reduction in toxicity. To date, gene therapy has avoided manipulation of germ line cells that would deliberately affect descendants of the treated individuals. The simplicity of the definition belies the practical difficulties which must be overcome before gene therapy can become a therapy capable of treating significant numbers of patients with common diseases.

Most gene therapy studies have focused on monogenetic disorders, whereby, if the normal gene product could be appropriately expressed at the relevant site, the abnormality could be corrected. The success of gene therapy relies on the development of gene-transfer systems or vectors that are safe, simple and efficient. Essentially, gene transfer involves the delivery, to target cells, of a therapeutic gene and sequences controlling its expression. The ideal route to manipulate tissues genetically is to transfer genes directly into the patient. Naked DNA alone, either in the form of an oligonucleotide or as a plasmid, is the simplest form of a gene-transfer reagent that can be used to transfect cells. However, naked nucleic acid is unstable in most tissues of the body and, as injecting purified plasmid DNA directly into cells requires individual cell manipulation, this is not a practical means of gene transfer *in vivo*. Encapsulating DNA within a lipid particle protects the nucleic acid from degradation by cellular enzymes and, although the plasmid–liposome complexes are still an inefficient method of delivering therapeutic genes to target cells, these vectors have been used in clinical trials of cancer and cystic fibrosis. Until improved *in vivo* methods are developed, the more indirect *ex vivo* manipulation will continue to dominate gene-transfer protocols. In the *ex vivo* approach, cells are removed from the patient, transfected with the therapeutic gene in culture and then the genetically corrected cells are subsequently returned to the individual.

5.2.2 Ex vivo *gene delivery*

Viruses have evolved to be extremely effective in transferring their genetic material into recipient cells and also at evading host-defence mechanisms. Viral vectors which have been engineered both for safety and to accept human genes have to date received the most attention as gene-transfer systems. Wild-type viruses can be immensely destructive but can be converted into safe gene-transfer vectors by substituting therapeutic genes for genes involved in viral

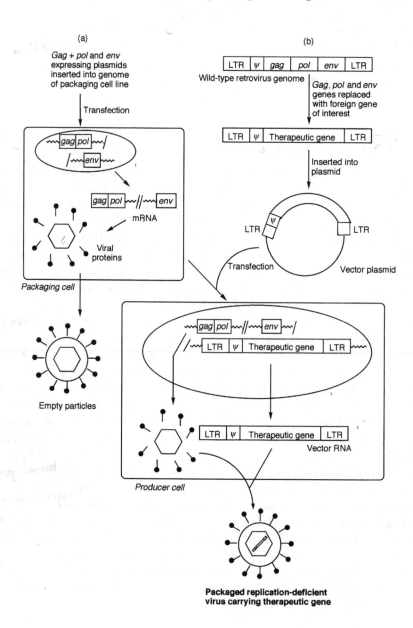

Packaged replication-deficient
virus carrying therapeutic gene

replication and virulence. Such replication-deficient viruses are therefore able to transfer genes to cells but are not able to multiply or cause disease. Viral-mediated gene-transfer methods result in the addition of a normal copy of defective genes to the cells that need correction rather than modifying mutant cells directly – they do not correct the underlying genetic defect that causes the disease.

The structure and life-cycle of retroviruses make them ideally suited to be gene-transfer vehicles. When a retrovirus enters a cell, the RNA genome is reverse-transcribed and the DNA product becomes stably integrated into the host chromosomal DNA and is expressed over extended periods (see Scheme 4.4 above). Efficient integration of the virus into the host cell requires cell replication, thus limiting the use of this vector to transferring genes into proliferating tissues. Nevertheless, the majority of gene-therapy trials to date have used retrovirus vectors.

A retroviral vector is composed of two parts, namely the modified viral genome and the structure, i.e. the virion which acts as the vehicle to introduce genes into cells. The wild-type viral genome consists of three structural genes, i.e. *gag*, *pol* and *env* which encode the capsid proteins, reverse transcriptase and the envelope glycoproteins, respectively, and a packaging signal (ψ) which recognizes the RNA transcript to be encapsidated. These genes are sandwiched between two long terminal repeat (LTR) sequences which are required for viral integration and have binding sites for regulatory proteins. The *gag*, *pol* and *env* genes are replaced with the foreign gene of interest and the resulting construct is inserted into a plasmid (Scheme 5.1). As RNA viruses replicate through a DNA intermediate, such manipulations can be conducted at the DNA level by using standard techniques of recombinant DNA. Since the

---

Scheme 5.1 Construction of a retroviral vector expressing a therapeutic transgene. (a) A packaging cell line is produced to help defective retroviruses to produce infectious vector particles. Plasmids encoding the retroviral structural genes *gag*, *pol* and *env* are introduced into a tissue cell line (most often NIH-3T3) by conventional calcium phosphate-mediated DNA transfection. The structural genes are under the control of the regulatory sequences within the LTR but without a packaging signal the genomic transcript cannot be encapsidated and therefore the cell-line produces 'empty' (i.e. genome-deficient) virions. The *gag–pol* and *env* genes are split into two separate transcriptional units as a safety feature so that in the presence of the retroviral vector in the producer cell an additional recombination event between these elements is, in principle, required to produce a wild-type virus. (b) A plasmid carrying the ψ sequence, a transgene in place of the *gag*, *pol* and *env* genes and flanked by the LTR sequences is transfected into the packaging cell. Integration of this plasmid into the genome of the packaging cell generates the producer cell. As it is only the RNA transcripts generated by the integrated plasmid-derived proviral sequences that contain a ψ packaging signal, then only vector RNA is packaged into virions. Thus the RNA generated by the vector construct combines with the constitutively expressed structural proteins of the packaging line to produce replication-incompetent, infectious retroviral particles carrying the therapeutic gene (LTR, long terminal repeat; ψ, packaging signal)

structural genes have been removed from the retrovirus, the vector is replication-deficient and requires 'help' to produce infectious viral particles. The viral replication functions are provided by packaging cells that are engineered to contain copies of *gag*, *pol* and *env* but lack the packaging signal so that no helper virus sequence becomes encapsidated and only empty virus particles are generated. These packaging cells are transfected with the retroviral vector plasmid which produces RNA that can be packaged and released as vector viral particles. Wild-type virus could be generated from the producer cell lines through recombination between the replication-deficient vector plasmid and the endogenous retroviral sequences. In order to reduce the likelihood of this occurring, the viral replication genes are split from each other so that two recombination events would then be necessary to produce a replication-competent virus.

Using the envelope proteins, the retroviral vector binds to proteins on the surface of the target cell and through a mechanism of receptor-mediated endocytosis becomes internalized (Scheme 5.2). In the cytoplasm, the reverse transcriptase derived from the *pol* gene converts the RNA genome of the virus into double-stranded DNA that is randomly inserted into one of the target cell's chromosomes. Once integrated into chromosomal DNA, the therapeutic gene is translated into its protein product. A retrovirus that can stably integrate a therapeutic gene into a patient's chromosomes where it will be expressed in perpetuity could be the ideal means of delivering a continuous supply of normal gene product needed to treat inherited diseases. In theory, the target cell's genome is permanently modified and the integrated gene could be expected to continue to express the gene product throughout the lifetime of the host cell and to be passed on to all progeny cells. In practice, this doesn't happen.

Retroviral vectors integrate at random within the genome of the target cell with the possible consequence of disrupting an essential gene or altering genes that favour cancer development. Activation of an oncogene, for example, will not necessarily lead to the formation of cancer or tumours as it is normally only one of a series of events required to generate a malignancy. However, insertional mutagenesis does raise the risk/benefit issue – is the increased risk of a subsequent tumour acceptable when compared to the potential benefit the gene therapy approach? A balance has to be achieved. A second safety issue is the possibility of generating infectious wild-type retrovirus in the vector preparations. Retroviral vectors cannot be made synthetically and must be produced by cultured cells and consequently stringent testing is necessary to ensure the absence of helper virus in vector preparations for use in humans.

Vectors derived from viruses other than retroviruses present their own advantages and disadvantages (Table 5.3). Those based on human adenoviruses are the most popular alternatives to retroviruses because they are safe; the wild-type virus causes nothing more serious than mild respiratory infections in humans. A major advantage of adenoviruses are their potential to carry large segments (approximately 7500 bases) of DNA and their ability to infect a large variety of cell types including non-dividing cells such as neurones

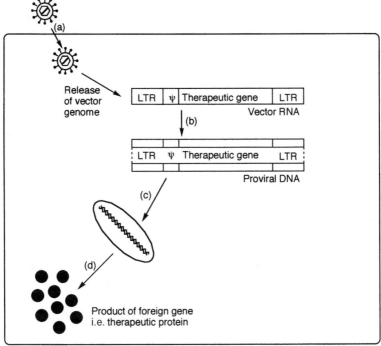

Scheme 5.2   Gene transfer. (a) The vector is delivered to the target cell by infection. The membrane of the viral vector fuses with the target cell allowing the RNA to enter. (b) The virally encoded enzyme reverse transcriptase converts vector RNA into RNA–DNA hybrid and then this is converted into double-stranded DNA. (c) The double-stranded viral DNA then enters the nucleus and integrates into the host chromosomal DNA. (d) The host-cell machinery transcribes and translates the transgene to make the foreign protein. As the viral genes have been replaced by a transgene, only the protein product of the transgene is made instead of new viral particles

Table 5.3   Gene transfer vehicles

Method	Advantages	Disadvantages
Retrovirus	• Chromosomal integration • Wide host cell range	• Random integration • Dividing cells only
Adenovirus	• No cell division required • Does not cause serious disease • Large capacity for foreign genes	• No integration • Transient expression • Immune response
Adeno-associated virus	• Chromosomal integration • Does not cause disease	• Small capacity for foreign genes
Non-viral	• Relatively simple • No immune response	• Very low efficiency • No integration

and muscle. One feature of adenoviruses which can be viewed either as an advantage or disadvantage is that the vector does not integrate into the host chromosomal DNA. This avoids possibly disturbing vital cellular genes or abetting cancer formation but gene expression is transient as the transferred genetic information remains extra-chromosomal and it may not be passed efficiently on to daughter cells. Transient expression of a gene without vector integration might be acceptable if short-term expression of the gene is sufficient for a clinical response, e.g. destruction of malignant cells. A more serious drawback to the use of adenovirus vectors in patients is their ability to evoke strong anti-vector immune responses. Although an initial round of treatment may generate high amounts of the desired proteins, the host's immune system may recognize and eliminate the vectors upon repeated administration, thus precluding continued therapeutic benefit.

The adeno-associated virus does not cause human disease and the viral DNA preferentially integrates into a specific region of chromosome 19 following infection. The major drawbacks to using the adeno-associated virus vector system for gene therapy are twofold. First, the entire viral genome is only about 5000 bases in length so that the amount of transgenic DNA that can be delivered is limited, and secondly, the virus requires co-infection with a helper virus such as the pathogenic adenovirus in order to replicate.

Non-viral vector-mediated systems are versatile and safe but they are substantially less efficient than viruses. The most promising of the non-viral DNA delivery systems uses synthetic cationic liposomes. The general principle is simple: positively charged cationic liposomes interact electrostatically with negatively charged DNA molecules to form complexes which are capable of entering a cell. Cationic liposomes are usually formed from a combination of a cytofectin, a positively charged amphiphilic molecule such as DOTMA (**5.1**) or the cholesterol derivative DC-Chol (**5.2**) and a neutral lipid, e.g. DOPE (**5.3**). Lipofectin, a 1 : 1 mixture of DOTMA and DOPE and DOPE/CD-Chol

(**5.1**) DOTMA

(**5.2**) DC-Chol

(5.3) DOPE

liposomes have been involved in cystic fibrosis gene-therapy clinical trials. While these liposomes could transfect airway passages, they were not efficient enough to bring about clinical benefit.

Transfection efficiency is low in practice because too few cells receive and express the exogenous DNA. The low efficiency of DNA delivery from outside the cells to inside the nucleus is a combination of slow uptake across the membrane, inadequate release of DNA molecules into the cytoplasm and inefficient trafficking to the nucleus. The liposome/DNA complexes are taken up by cells usually through endocytosis and while this is an efficient, albeit slow, process a large portion of delivered nucleic acid is trapped in endosome compartments. When DNA does succeed in dissociating from the complex and diffuse into the cytoplasm it is in an unfriendly environment and can be degraded by nucleases. Finding the nucleus is the final obstacle for DNA delivery, and in this respect, synthetic systems are more inefficient than viral vectors. The amount of liposome currently required to deliver therapeutically useful levels of DNA is too large to be clinically useful.

Viruses have evolved a specific machinery to deliver DNA to cells. They are able to totally escape and/or bypass endocytosis with extremely high efficiency and it is probable that viral components will be included into future DNA synthetic delivery systems.

As the human applications of gene transfer are broad, it is likely that no single vector system will be appropriate for all disorders. Questions that need to be addressed in developing a vector for human gene therapy include the following: Which cells constitute the target? Is the treatment likely to be *in vivo* or *ex vivo*? Is the requirement for gene expression temporary or permanent? – and all this without causing an immune response by the host to the introduced agent. Once designed, a vector needs to be tested for efficacy of gene transfer and efficacy of gene expression, the duration of gene expression, the ability for repeat dosing and the ability to target appropriate cells and avoid inappropriate cells.

5.2.3   Clinical applications

Single-gene defects

In clotting factor disorders, almost any somatic cell could be used as a target for gene therapy because the gene product is secreted into the circulation and required systemically. In other cases, the gene must be delivered to a specific cell type. A number of diseases such as adenosine deaminase (ADA) deficiency and β-thalassaemia specifically affect the haematopoietic system. Haematopoietic stem cells are long-term repopulating cells that are present in small numbers in bone marrow and give rise to all types of terminally differentiated blood cells over prolonged periods. Gene transfer into a small number of haematopoietic stem cells could therefore produce a continuous supply of fully functioning cells and correction of these genetic diseases.

Much of the pioneering research into human gene transfer has focused on ADA deficiency – an inherited severe combined immunodeficiency (SCID) disorder. ADA is a housekeeping enzyme of purine degradation that catalyses the hydrolytic deamination of adenosine and deoxyadenosine to inosine and deoxyinosine (Scheme 5.3). A mutation in the gene encoding the ADA enzyme leads to the accumulation of large amounts of adenosine and deoxyadenosine which are toxic to T cells. Affected children are unable to generate normal immune responses; a condition which is fatal unless treated. The disease can be cured by transplantation of bone marrow from a fully MHC-matched donor but if this is not possible the levels of toxic metabolites can be reduced by exogenous enzyme replacement with bovine-ADA which has been conjugated to PEG to increase its half-life. Although PEG–ADA treatment reduces the systemic concentration of adenosine metabolites and results in a marked improvement in the disease, producing ADA within T cells gives a more complete correction of the disease.

The first human gene therapy clinical trial started in 1990 and involved treating two children with ADA deficiency by transferring the *ADA* gene into their T cells, i.e. the cells that are most affected by this disorder. The two girls, in whom response to replacement therapy had been incomplete, were treated *ex vivo* by infusion of their own cells with a retroviral vector that expressed human ADA. The protocol involved taking peripheral blood from the patients, separating the T cells by apheresis and then culturing the lymphocytes in the presence of IL-2 and muronomab-CD3 to stimulate proliferation. The cells were then transduced with the *ADA*-containing retroviral vector, further expanded for 9–12 days and then reinfused back into the patients at

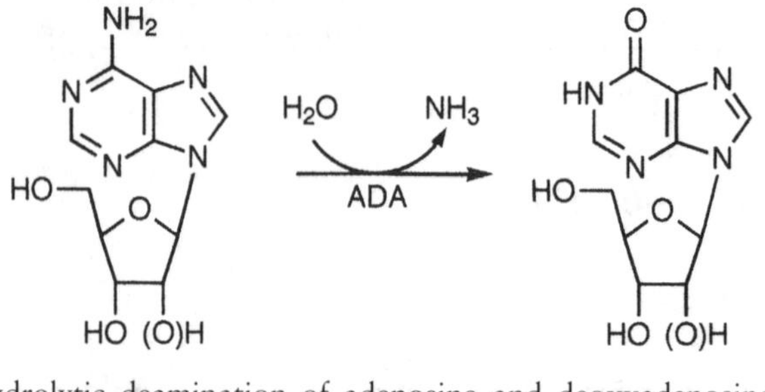

Scheme 5.3 Hydrolytic deamination of adenosine and deoxyadenosine to inosine and deoxyinosine

six-to-eight week intervals over a one-to-two year period. Within five to six months of beginning gene therapy the peripheral T-cell count of the first girl rapidly increased in number and stabilized in the normal range and ADA levels increased to approximately one-quarter of normal. Her immune system recovered as shown by antibody responses to vaccines to tetanus toxoid and *Haemophilus influenzae* type b (Hib). Even after the gene treatment ended, the vector and ADA gene expression, somewhat unexpectedly, persisted in transduced cells and their descendants. The vigour of the girl's immune responses has gradually diminished in the years following gene therapy but it still remains in the normal range, albeit at the low end. Treatment was less successful in the second patient as gene transfer efficiency was low, thus highlighting one shortcoming of the technique, namely individuals' response to treatment is variable and results may be inconsistent. Throughout the trial both patients were receiving PEG–ADA enzyme and so it was difficult to quantitate by how much the immune function was improved by the gene therapy alone.

In the above study, the *ADA* cDNA and a neomycin gene (*neo*) were inserted into the Moloney murine leukaemia virus (MoMuLV), the most widely used virus for retrovirus vectors intended for clinical use. Incorporation of the antibiotic resistance gene *neo* into the vector is essentially a marker to aid in the identification of genetically modified cells and their progeny in the circulation of each patient. Transcriptional control of the *ADA* gene was regulated by retroviral LTR sequences, while the *neo* gene was controlled by an internal promoter. Although retroviral vectors are in essence relatively simple, containing the 5′ and 3′ LTRs, a packaging sequence and a transcription unit composed of the gene or genes of interest, various configurations of the vector can be generated. Choice of virus backbone, position of genes relative to viral splice sites, whether to include a selectable marker and/or incorporate internal promoter sequences, are just some of the factors to be considered in vector design. Choice of target tissue is also a variable in gene therapy. Bone marrow stem cells, rather than the terminally differentiated T cells which have a finite and relatively short life-span, are a better target for long-term reconstitution of the immune system and amelioration of ADA. The frequency of haematopoietic stem cells is, however very low in bone marrow and gene transfer is inefficient. Since the original clinical trial in 1990, the *ADA* gene has been transferred into bone marrow cells but with a vector configured to improve the expression of retrovirally transduced genes. In this second trial, bone marrow and circulating T lymphocytes of two children undergoing exogenous enzyme replacement therapy were both transduced with an *ADA* vector. Two different vectors were used to infect the two different populations of cells; the vectors were identical except for the presence of an alternative restriction site so that vector integrants in the recipient cells could be distinguished from each other. Within six months of beginning treatment, the T-cell repertoire normalized and both cellular and antibody immunity was restored to both patients. Peripheral T-cell-derived *ADA-*

transduced cells predominated during the two-year course of treatment but after discontinuation of gene therapy these were progressively replaced by bone marrow-derived T cells suggesting successful gene transfer into long-lasting progenitor cells. The procedure is furthermore safe; no malignancy arising from insertional mutagenesis associated with retroviral-mediated transfer has been detected in any of the gene therapy-treated ADA patients to date.

The results from the two trials could be open to interpretation as all of the children received the standard therapy of PEG–ADA. Enzyme replacement treatment in these gene therapy trials was a two-edged sword; stopping PEG–ADA may have exposed the patients to recurrent immunodeficiency, which was unacceptable, but the very presence of exogenous enzyme may have made it harder for cells containing the new gene to thrive.

The ethical dilemma of whether to withhold a partially corrective treatment during gene therapy was not a complication in a recent clinical trial of two boys with severe combined immunodeficiency-XI (SCID-XI). The latter is an X-linked inherited disorder caused by mutations of the gene encoding the γc cytokine receptor sub-unit of several of the interleukin receptors. Deficiency in the interleukin receptors blocks T and NK lymphocyte differentiation and children affected by this disorder are forced to live within sterile bubbles to avoid any threats to their non-existent immune systems. The disease can be cured by bone marrow transplantation but this has been the only treatment – there is no exogenous protein replacement therapy. The gene therapy trial again used a Moloney-derived retroviral vector to deliver the therapeutic gene *ex vivo* to haematopoietic stem cells of the SCID-XI boys aged 11 and 8 months. Levels of gene transduction were much higher than those of the earlier ADA studies and this was attributed to the inclusion of a factor that greatly enhanced stem cell growth in culture and the use of fibronectin-coated culture vessels. At the time of writing (May 2001), 10 months after receiving transduced cells, the numbers of T, B and NK cells of the immune system were normal as were a number of measures of immune function, such as specific responses to antigens. Both children were able to leave protective isolation in hospital after three months and have been at home ever since. Long-term follow up will be necessary, as in the case of the ADA patients, but the evidence is amassing to conclude that *ex vivo* gene therapy can be a safe and effective addition to the management of some patients with immunodeficiency disorders.

Only those genetic diseases which affect the haemopoietic system will be corrected by gene transfer into bone marrow stem cells. Lung diseases, e.g. cystic fibrosis and emphysema, can potentially be corrected by gene transfer into cells of the respiratory tract.

Cystic fibrosis (CF), one of the most common monogenetic recessive hereditary diseases affecting people of northern European descent, is caused by mutations to the gene encoding the CFTR (cystic fibrosis transmembrane conductance regulator) channel. The CFTR protein is an adenosine 3′,5′-

monophosphate (cAMP) regulatable chloride channel in the apical epithelium. More than 400 mutations in the *CFTR* can occur. The commonest form of CF, accounting for approximately 70% of cases overall, is a three base-pair deletion resulting in the loss of a phenylalanine residue in one of the ATP binding regions. Aberrant chloride secretion due to defective protein production initiates a pathophysiological cascade. Chloride secretion controls the volume and hydration of airway mucus and failure of chloride transport causes the accumulation of viscous mucus in the epithelial cells lining the airways of the lung. As the airways become blocked there is difficulty in breathing and the lungs are subject to repeated and persistent infection. Pancreatic function is also affected so that there is poor digestion of fats and proteins but lung disease predominates and death results from progressive respiratory failure in greater than 95% of CF patients (median survival age is 29 years). Most CF patients are currently treated by a combination of chest physiotherapy to drain the air passages of mucus, antibiotics to keep the lungs clear of infection and enzyme supplementation to compensate for pancreatic disease. The best-tested method for correcting the basic defect in CF is lung or heart–lung transplantation. Restoration of normal cAMP-regulated chloride conductance could also be achieved by complementation with a normal gene.

As the respiratory epithelium is the site of the most severe pathophysiology and morbidity this is the tissue that must be corrected. Respiratory epithelium cells are, however, difficult to harvest and reimplant and therefore most gene-transfer studies have focused on direct *in vivo* delivery of vectors to the airways of patients with CF. Adenovirus vectors are particularly suitable for airway epithelial cells as the virus is a common cause of upper respiratory tract infections.

The first generation of adenovirus vectors (Scheme 5.4) were produced by deleting the E1 gene from the adenoviral DNA and inserting the cDNA for human *CFTR* in its place. These vectors were then packaged in a cell line that provided the essential regulatory E1 gene. In clinical trials in CF patients using nasal administration of the vector, transient gene expression (one to two weeks) was observed but some patients developed an inflammatory response. The induction of inflammatory responses and the inability to re-administer the vectors due to the generation of antibody responses are major problems with the use of adenoviral vectors. An additional problem in 'mature' patients is that gene-delivery vectors cannot cross accumulated mucus. Second-generation vectors are being developed in which deletion of the E1 gene is coupled with the deletion of other essential early genes. The intention is to further cripple the ability of the virus to express viral proteins which cause immune responses and thus circumvent the clinical toxicity observed with the first-generation vectors. The safety of the second generation constructs has yet to be confirmed in clinical trials. Viral systems are not the only potential means for delivering the *CFTR* gene into cells and wrapping up *CFTR* cDNA in liposomes is an alternative method of transfer which is also being explored.

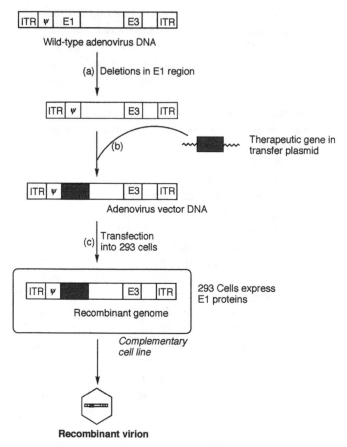

Scheme 5.4 Construction of an adenovirus vector expressing a therapeutic transgene. The wild-type adenovirus genome is divided into early (E1 to E4) genes with regulatory functions and late (L1 to L5) genes that code for structural proteins. Adenoviruses are converted into vectors for gene transfer by (a) deleting the E1 gene (responsible for transactivation of the other viral genes) and, if space is needed, the non-essential E3 region, and (b) inserting the therapeutic gene plus regulatory sequences. (c) The E1-replication-defective chimeric construct is transfected into a cell line, commonly the human embryonic kidney cell line 293, which constitutively expresses the products of the E1 gene. The transferred genetic information functions in an extrachromosomal fashion to direct the expression of its product. While the non-integrating nature of the vector is an advantage in that it poses little risk of insertional mutagensis, it is disadvantageous in that expression of the therapeutic gene may be transient (ITR, inverted terminal repeat; ψ, packaging signal)

Cancer

As gene transfer is in effect a form of drug delivery, it has the potential to ameliorate cancer and other acquired disorders that have a genetic component. More than 50% of all current clinical gene therapy research is focused in cancer. Experimental cancer gene therapy takes varied forms.

Immunotherapy (vaccine therapy) – activating and focusing of the immune responses on to the tumour (see also Section 4.5.2 earlier)   Immunity is a systemic reaction and therefore it has the potential to eliminate all cancerous cells in a patient's body. The problem is that human tumours seem to be predominantly weakly immunogenic. In immunotherapy, the goal is to enhance the body's own immune defences against cancer. One of the most widely tested gene-transfer approaches at the moment aims to boost the T_h cell response by modifying a patient's cancer cells with genes encoding cytokines. The key concept underlying the use of such cytokine gene-transduced tumour vaccines is that the cytokine is produced at very high concentrations local to the tumour and should therefore result in significant antitumour immune responses without the toxicities that can be associated with systemic cytokine administration. Tumour-infiltrating lymphocytes are isolated from a patient, cultured and then transduced *in vitro* with the cytokine viral vector. The altered tumour cells are then returned to the patient where they should secrete the cytokine and thereby elicit vigorous cellular immune activity at the site of the tumour. A variety of, predominantly retroviral, vectors have been constructed to contain cytokine genes and a large number of phase 1 clinical trials to test the efficacy of genetically modified tumour vaccines using cytokines such as IL-2, tumour necrosis factor (TNF), γ-interferon (IFN-γ) and granulocyte-macrophage colony stimulating factor (GM-CSF) are currently underway. The experimental protocols are often carried out on patients with terminal cancer who have previously received intensive conventional anticancer therapy. This weakens their immune system and therefore if vaccines do activate immunity in these individuals the responses may not be easily noticed. Preliminary results from the early trials do provide convincing evidence that vaccinations with cytokine gene-transduced autologous tumour cells can induce enhanced tumour-specific T and B cell immunogenicity. Whether this approach will be sufficiently potent to induce clinically meaningful responses in many patients with advanced stages of cancer remains to be seen.

Suicide gene therapy – rendering cancer cells highly sensitive to selected drugs   The rationale behind this approach is to impart cancer cells with genes that give rise to toxic molecules that can kill the tumour. Human thymidine kinase phosphorylates thymidine and a limited number of structurally related analogues, whereas herpes simplex virus thymidine kinase (HSV-TK) catalyses the phosphorylation of a variety of abnormal nucleosides such as the drug ganciclovir (**5.4**). Ganciclovir lacks a 3′-hydroxyl group and acts as a chain terminator when incorporated into DNA. In theory, transducing tumour cells with a vector containing the *HSV-TK* gene will phosphorylate ganciclovir, whereupon the ganciclovir triphosphate then blocks the DNA synthesis machinery and kills the cell. This method is in clinical trials for the treatment of glioblastoma multiforma, a malignant brain tumour. As the only dividing cells in the area of a growing brain tumour are the tumour cells and

(5.4) Ganciclovir

cells of the vasculature supplying blood to the tumour and, as retroviral vectors only transduce dividing cells, the only cells to receive the vector should be those of the tumour and its blood vessels. Mouse producer cells making retroviral particles carrying the *HSV-TK* gene, and a neomycin-resistance gene as a marker, were inoculated into the tumour masses of 15 patients. After seven days, the patients were treated with ganciclovir. About one-quarter of the patients responded to the treatment. *In situ* hybridization with antisense probes to *HSV-TK* indicated that vector-producing cells were surviving at seven days but that gene transfer to the tumour cells was limited. The observed reduction in tumour size was consequently not wholly due to the direct effect of the phosphorylated ganciclovir on the transduced tumour cells. An indirect mechanism (the bystander effect) was contributing significantly to the anti-tumour activity – toxic ganciclovir triphosphate generated in one gene-modified cell was spreading to neighbouring cells through gap junctions and killing them.

Tumour suppressor gene – replacement of a lost or damaged cancer blocking gene   One of the most commonly mutated tumour suppressor genes in human cancer is *p53*, sometimes known as the 'guardian of the genome'. The *p53* gene product links three cellular functions, i.e. proliferation, DNA repair and death. If the DNA is flawed, the p53 protein may halt cell division until the damage is fixed or it may induce apoptosis. Over 2000 *p53* mutations have been identified and the codon and the type of mutation varies according to cancer type. Specific base changes often result in stabilization and accumu-lation of p53 mutant proteins within the cell. Mis-sense mutations can therefore give rise to new tumour-specific peptide sequences which can act as targets for T cell-mediated immunotherapy. In addition, p53 inactivation may also arise, not because the gene has changed, but because other proteins, e.g. the E6 of the human papillomavirus, bind and block transcriptional activity. The importance of p53 in cell function makes it an obvious target for gene therapy. Adenovirus vectors are the most frequently used gene-delivery systems to effect transfer of *p53* to cancer cells but expression lasts for only a short time. The immune system is behind the short-term expression as antibodies to adenoviral proteins are generated so that on repeat administra-tion of the recombinant vector the virus is destroyed before it can deliver genes to tumours (cf. HAMA response in murine monoclonal antibody therapy, see Section 4.2.2 earlier). Preclinical experiments have, however, demonstrated that restoration of wild-type *p53* function in the cancer cell by gene transfer is sufficient to cause antitumour effects such as cell-cycle arrest and induction of apoptosis. Evidence from *in vitro* and *in vivo* preclinical studies have also indicated that tumour cells having a wild-type *p53* status are more sensitive to chemotherapeutic agents and radiation than cells that lack functional *p53*. In theory, a major limitation to using gene transfer to activate tumour suppressor genes is that the corrective gene must be delivered to every tumour cell – an impossibility – or otherwise the unaccessed cells will continue to grow uncontrollably. However, there have been reports of a bystander effect in wild-type *p53* gene therapy whereby transduced tumour cells can mediate the killing of non-transduced tumour cells and this might help correct more tumour cells. Therefore, a treatment strategy that could result in a bystander effect offers tremendous therapeutic advantages. Gene transfer of *p53* into cancer cells has entered clinical testing. Trials are at phase 1 stage and evaluate the safety and antitumour efficacy of administrating adenoviral vectors expressing wild-type *p53* to patients with, mainly, lung or head and neck carcinomas. Information to date suggests that intratumoural injection of the *p53* vector is feasible, well tolerated and can mediate apoptosis in human tumours *in vivo*, as has been shown *in vitro* and in animal models.

While it is still early days for cancer gene therapy. there are grounds for optimism. Techniques to improve therapeutic efficacy such as optimization of gene delivery and enhancement and sustainment of gene expression are needed before gene therapy becomes an established treatment option for cancer.

5.3 Antisense Therapy

5.3.1 Introduction

Diseases characterized by an abnormal protein or a surfeit of a normal protein can potentially be treated by selectively blocking the disease-causing genes. When a gene is expressed, the base sequence of DNA is transcribed into single-stranded RNA and then translated into a protein. The basic idea of antisense oligonucleotide (ASO)-based therapy is to interrupt the flow of genetic information from gene to protein by using synthetic oligonucleotides targeted to specific sequences in the mRNA (Figure 5.1).

There are two possible mechanisms by which the translation of mRNA can be blocked by the binding of a complementary (and hence antisense) oligo-nucleotide: (i) by base-specific hybridization, thus preventing access by the translation machinery, i.e. hybridization arrest, or (ii) by cleavage of the RNA strand of the RNA–DNA hybrid by the cellular enzyme ribonuclease H (RNase H). Ribozymes (RNA enzymes) are another type of antisense agent

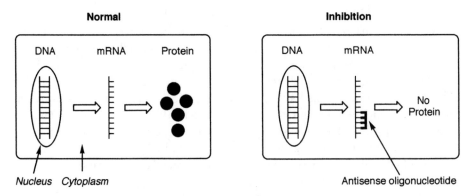

Figure 5.1   Inhibition of mRNA function with antisense oligonucleotides

and these are discussed below in Section 5.4. Gene expression can also be disrupted at the transcription level by inserting an oligonucleotide sequence into the major groove of the DNA double helix and formation of side-to-side hydrogen bonds (Hoogsteen base pairs) with one of the strands (referred to as triplex technology or antigene approach). The resulting triple helix cannot be transcribed.

The allure of the antisense strategy lies in its potential specificity – the hybridization interactions that depend on hydrogen bonding (Crick–Watson or other types of base-pairing) and the defined sequence of the nucleic acid means that if an oligonucleotide can selectively target just the pathogenic gene then there will be no toxicity associated with the procedure. Furthermore, the known sequences of nucleic acids facilitate the design of antisense oligonucleotides. This then precludes the need for either laborious hit-and-miss design or the need to synthesize and screen many candidate compounds as is often the case with classical drug compounds. Scheme 5.5 outlines how an antisense agent targeted to the first 15 bases of c-*myc* gene could be developed.

During the process of gene transcription mRNA is made from the complementary chain of DNA – the template strand. The coding region of the c-*myc* gene starts at nucleotide 16 of exon 2 and the first 15 bases, ATG CCC CTC AAC GTT corresponds to the N-terminal pentapeptide Met-Pro-Leu-Asn-Val of the translated protein. The mRNA has the same sequence as the DNA coding strand except that thymine (T) residues are replaced by uracil (U). Furthermore in the transcription process the non-coding regions, the introns, are removed from the primary transcript and the mature mRNA is composed only of the coding regions or exons. The antisense DNA sequence to the mRNA is derived by complementary Crick and Watson base pairing, i.e. C binds to G and A (or U in the mRNA) with T. As oligonucleotides are conventionally written with the 5' base on the left and the 3' on the right then the antisense sequence against the initiation site of the c-*myc* gene is $^{5'}$ AAC

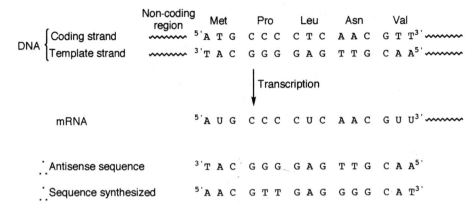

Scheme 5.5 Design of an antisense agent for the c-*myc* gene. By convention, when a DNA sequence is specified it is presented in the coding strand

GTT GAG GGG CAT $^{3'}$. This sequence would not be selective and would inhibit c-*myc* expression in all cells. For more selective inhibition it would be better to choose a target sequence unique to a particular disease. In Burkitt's lymphoma translocation of the c-*myc* gene creates abnormal mRNA molecules which retain normally spliced intron sequences. An ASO complementary to the abnormal sequence present in tumour cell RNA might be more effective in controlling the proliferation of tumour cells without affecting normal cells.

ASO-based therapy is in its infancy and most studies have focused on cancer and viral infections. Overexpression of specific genes have been correlated as a causative factor in some cancers, e.g. c-*myc* with several cancers of the lung, colon and neuroblastomas, and *HER2/neu* with breast cancer. Down-regulating or specifically turning off the expression of oncogenes therefore offers the possibility of selectively ablating tumours without the systemic side-effects often associated with conventional antineoplastic drugs.

Retroviruses have formed a major target in studies of antiviral oligonucleotide strategies because of the world-wide AIDS problem. In the normal replication mechanism, reverse-transcribed copies of the retrovirus are integrated into the chromosomes of the host cell and are expressed by the standard transcriptional and translational mechanisms. Targeting mRNA of viral-specific gene products, e.g. Gag protein, may prevent viral replication and production, thereby stopping the disease. Other targets could include splice-site junctions of pre-mRNA i.e. specific sequences required for excision of introns, thus blocking production of mature mRNA or the packaging or protease cleavage sites of the mature mRNA.

ASO therapy will not be a universal cure-all and diseases characterized by a lack of gene product will not be amenable to this approach.

5.3.2 Development of an antisense oligonucleotide

From a molecular point of view, ASOs are the logical approach for the treatment of any diseases caused by aberrant gene expression. ASOs are drugs for the future and several hurdles need to be overcome before these molecules become adopted for clinical use. In order to be an effective antisense agent the oligonucleotide should meet the following criteria:

(1) be synthesized easily and in bulk;

(2) be stable against nucleases in cells and body fluids;

(3) be efficiently taken up and retained by target cells;

(4) have a strong and selective affinity toward the target sequence.

(1) Oligonucleotide synthesis

Chemical synthesis is the driving force behind the development of oligonucleotides (strictly speaking, oligodeoxyribonucleotides) as therapeutic agents. Oligonucleotides are usually prepared by the sequential coupling of nucleoside monomers to a terminal nucleoside attached to a solid support. The advantages and drawbacks of solid-phase oligonucleotide synthesis are similar to those experienced in peptide synthesis (see Section 2.3.1 above); chain assembly is rapid and can be automated but all purification and characterization of the macromolecule is carried out after cleavage from the solid support. The crux of the synthesis is formation of the internucleoside $3'-5'$-phosphodiester linkage. Coupling of two nucleosides specifically to form a $3'-5'$ internucleotide link only occurs if all nucleophilic centres not involved in the linkage are protected. As with peptide synthesis, this requires the correct combination of temporary (i.e. removed after each coupling step) and permanent (i.e. remains attached to the oligonucleotide during chain assembly and only removed at the end of the synthesis) protecting groups. The exocyclic amino groups of the heterocyclic bases are permanently protected with base-labile acyl groups, i.e. benzoyl for adenosine and cytidine, and isobutyryl for guanosine (Figure 5.2). An acid-labile temporary dimethoxytrityl group (DMTr, 5.5) is used to protect the $5'$-OH function of the deoxyribose ring.

   The four essential steps in the solid-phase synthesis of oligonucleotides are as follows

(1) attachment of the first nucleoside to the support;

(2) assembly of the oligonucleotide chain;

(3) deprotection and removal of oligonucleotide from the support;

(4) purification and characterization.

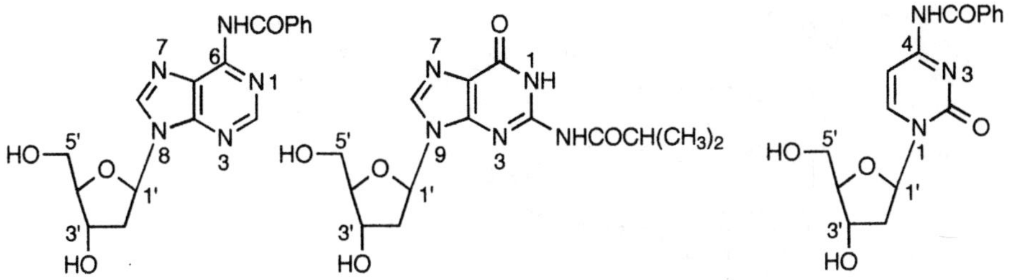

N^6-Benzoyl deoxyadenosine N^2-Isobutyryl deoxyguanosine N^4-Benzoyl deoxycytidine

Figure 5.2 Protected deoxynucleosides; note that no protection is necessary for deoxythymidine since it does not have an exocyclic amino group

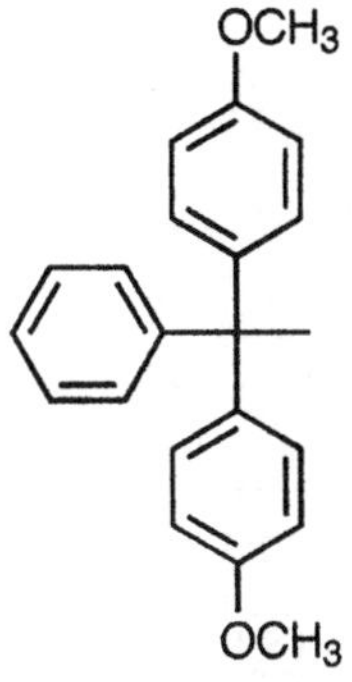

(5.5) Dimethoxytrityl (DMTr)

Oligonucleotides are synthesized from the 3'-end to the 5'-end with 3'-terminal nucleoside attached by a succinate linkage through its 3'-OH group to the inert insoluble support. Each nucleoside addition involves removal of the acid-labile DMTr group and coupling of the activated monomer plus washing and any ancillary steps such as capping and oxidation.

There are three methods of making an internucleotide bond between two nucleoside building blocks, i.e. involving the phosphotriester, phosphoramidite and H-phosphonate intermediates. Phosphoramidite (also known as phosphite triester) chemistry is currently the most efficient method of assembling oligonucleotide chains. It entails reacting the 5'-OH of the support-bound nucleoside **5.6** or growing chain with a protected 3'-O-phosphoramidite nucleoside **5.7** in the presence of tetrazole, a mild acid catalyst (Scheme 5.6(a)). Protonation at nitrogen converts the phosphoramidite into a highly reactive phosphitylating agent and the product of the coupling is the phosphite triester **5.8** which is then immediately oxidized with iodine to the more stable 3'–5' phosphotriester linkage **5.9**. The internucleoside phosphate carries a

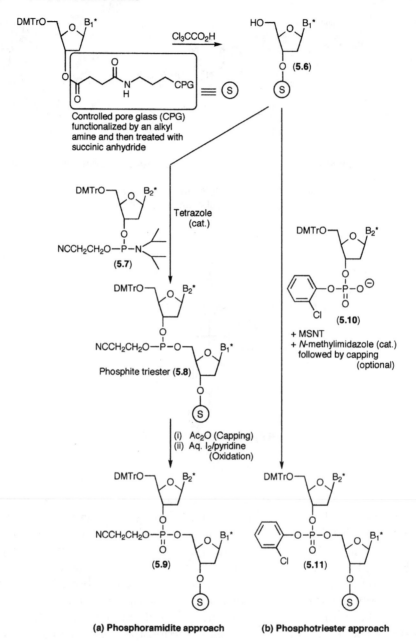

Scheme 5.6 Formation of an internucleotide bond by (a) phosphoramidite and (b) phosphotriester chemistry (B*, protected base)

protecting group and this is removed once chain assembly is complete. To extend the chain, the DMTr group is removed by treatment with acid to liberate the 5′-hydroxyl group ready for further coupling.

Coupling a 3′-phosphodiester nucleoside unit **5.10** with the 5′-OH of the DNA chain yields the phosphotriester product **5.11** in one step (Scheme 5.6(b)). Condensation is brought about by mesitylenesulphonyl 3-nitro-1,2,4-triazolide (MSNT). The two coupling processes yield phosphotriesters which differ in only their phosphate protecting group.

In the H-phosphonate mute the 5′-OH of the chain (**5.6**) reacts with a nucleoside 3′-H-phosphonate **5.12** (Scheme 5.7). Activation is achieved with a sterically hindered acid chloride such as pivaloyl chloride. The resulting H-phosphonate diester **5.13** is stable under further chain-extension conditions and oxidation of all phosphorous centres is carried out simultaneously at the end of the synthesis, in contrast to the amidite method.

By using automated synthesizers the assembly of protected oligonucleotides can be carried out on 0.2–1000 μmol scale (corresponding to a yield of approximately 1 mg up to 3 g of a 20-mer oligonucleotide after purification). Cycle times are roughly 8 minutes in length and thus a 20-mer can be assembled in just a few hours.

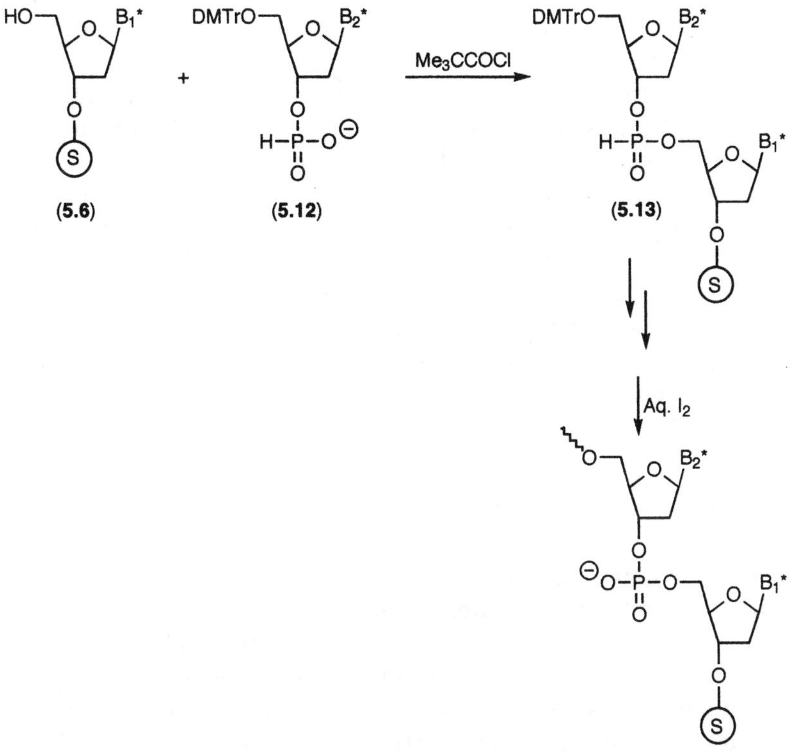

Scheme 5.7   The H-phosphonate chain assembly strategy (B*, protected base)

After the appropriate number of reaction cycles to extend the chain to the required length, the oligonucleotide is cleaved from the support and protecting groups are removed. The heterocyclic base protecting groups, the succinate ester linkage and, in phosphoramidite synthesis, the 2-cyanoethyl group, are all base-labile and are removed with aqueous ammonia. An additional deprotection step using *syn*-2-nitrobenzaldoximate ion or 2-pyridine carbaldoximate ion is needed in phosphotriester syntheses to cleave the 2-chlorophenyl protecting group. The crude synthetic oligonucleotides are purified by high performance liquid chromatography or polyacrylamide gel electrophoresis.

The phosphotriester method dominated the preparation of oligonucleotides for a long time, although the phosphoramidite chemistry gives a higher coupling efficiency (greater than 99% compared with 98%) and with fewer, if any, byproducts this assembly procedure is now frequently the preferred synthetic route.

(2)   Nuclease resistance

Targets for ASOs are intracellular and therefore the oligonucleotide must be relatively stable inside and outside of the cell and must be able to traverse the cellular membrane. Natural ASOs consist of phosphodiester oligomers but the internucleotidic phosphate linkage is very sensitive to digestion by nucleases present in extracellular fluids and in the intracellular environment. The half-life of a phosphodiester typically ranges from 15 to 60 min in sera and unmodified oligomers will probably never become pharmaceuticals. In order to extend the biological half-life and to improve uptake by cells, oligomers with a modified internucleotide phosphate linkage are prepared. While a variety of backbone modifications can be envisaged (Figure 5.3), the phosphorothioate and methylphosphonate are the most extensively investigated analogues. The phosphoramidite and H-phosphonate routes can both be adapted to prepare sequences with phosphorothioate and methylphosphonate internucleotide bridges.

Phosphorothioate oligonucleotides

In phosphorothioate oligonucleotides, one of the oxygen atoms not involved in the bridge is replaced by a sulphur atom. Sulphur can be introduced at any internucleotidic linkage of choice using oxidation of the phosphite intermediate with elemental sulphur during phosphoramidite chemistry (Scheme 5.8). If KSeCN is used for oxidation in place of sulphur, phosphoroselenide analogues can be obtained. An alternative method of synthesis, which relies on H-phosphonate chemistry, can introduce phosphorothioates non-selectively at all

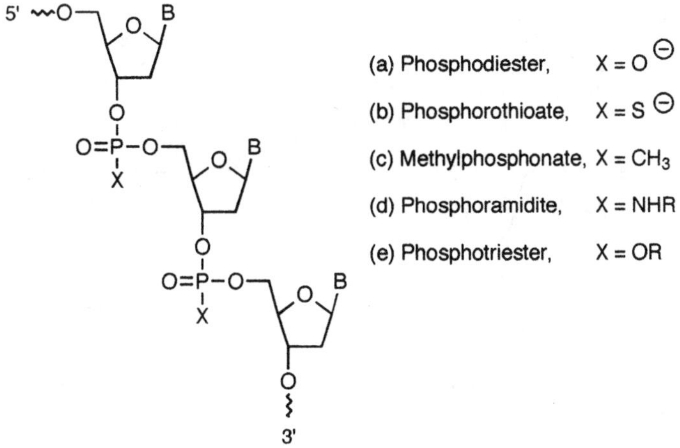

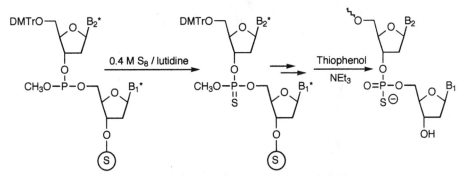

Figure 5.3 Oligonucleotide backbone modifications

linkages. Oxidation with sulphur is necessary only once, after chain assembly is complete (Scheme 5.9).

The substitution by sulphur for one of the non-bridging oxygen atoms at phosphorus produces a compound that retains its net charge and aqueous solubility. The substitution does, however, introduce chirality at the phosphorus atom; each phosphorothioate linkage can occur as either R_P or S_P diastereoisomers. The diastereoisomeric mixtures may potentially have very different biochemical, biophysical and biological properties compared with their 'pure' stereoregular counterparts. Stereoselective synthesis of oligonucleotide phosphorothioates yields either R_P- or S_P-rich populations but not fully stereoregular oligomers. The chirality problem can be avoided by substituting sulphur for both of the non-bridging oxygen atoms in the phosphodiester backbone. Condensing a phosphorothioamidite with a

Scheme 5.8 Oligonucleotide phosphorothioate synthesis by a modification of the phosphoramidite approach

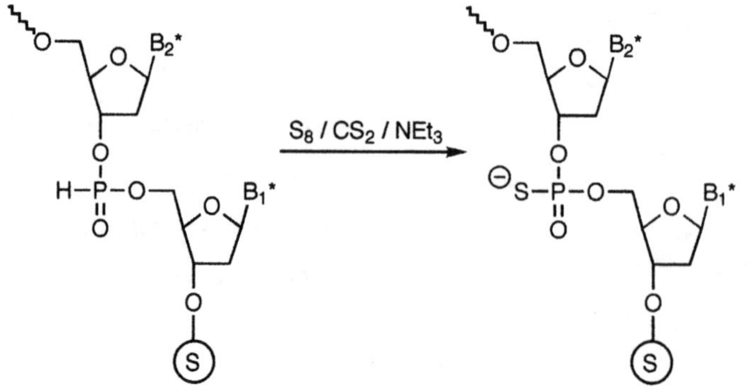

Scheme 5.9 Oligonucleotide phosphorothioate synthesis by a modification of the H-phosphonate method

second nucleoside with tetrazole catalysis and oxidizing the resulting inter-mediate yields the achiral phosphorodithioate.

Methylphosphonate oligonucleotides

Substituting a methyl group for one of the non-bridging oxygen atoms gives the methylphosphonate analogue. The methylphosphonate linkage can be introduced by using 3'-O-methylphosphoramidite nucleosides under similar conditions to standard solid-phase phosphoramidite synthesis. Nucleotides with only one or with several methylphosphonate bridges at any desired point in the molecule can be prepared by using the usual synthetic cycle

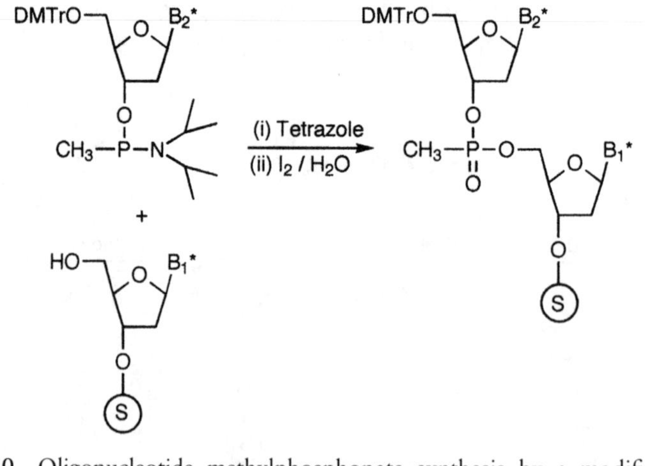

Scheme 5.10 Oligonucleotide methylphosphonate synthesis by a modification of the phosphoramidite method

(Scheme 5.10). The methylphosphonate bridge is more base-labile than the natural internucleotide linkage and milder conditions are necessary for deprotection and cleavage of the oligomer from the support (aq. NH_3 at room temperature for 2 h followed by ethylenendiamine–ethanol (1 : 1) for 7 h, cf. aq. NH_3 at 60°C for 8 h for natural phosphodiester oligonucleotides). The methylphosphonates are, like the phosphorothioates, diastereoisomeric at each modified phosphate.

The phosphorothioates retain the negative charge and are consequently still substrates for RNase H but they are degraded more slowly and have half-lives of greater than 24 h. The methylphosphonates are non-ionic analogues of natural phosphodiesters and are therefore stable to nucleases. In *in vivo* assay systems, both classes of analogue are found to be lost by excretion rather than by degradation.

(3) Cellular uptake

Most ASOs have been designed with the aim of inhibiting translation. The protein biosynthesis apparatus of the cell is located in the cytoplasm and in order for ASOs to be effective and to stop translation by hybridization they must pass through the plasma membranes into the interior of the cell. Methylphosphonates are uncharged and lipophilic and are thought to be taken up by cells in tissue culture via passive diffusion. Phosphorothioates are negatively charged and are slightly more lipophilic than phosphodiesters because of the presence of sulphur. They enter cells through receptor-mediated endocytosis after binding to cell surface proteins, although uptake is slower than for normal oligonucleotides. Most cells carry the receptor for charged oligonucleotides and neutral oligomers enter all cell types. Both classes of modified oligonucleotide are active in, usually, the µM range in cell assay but activity should be in the nM range for therapeutic effectiveness. Cellular uptake and efficacy of oligomers can be enhanced by encapsulation into liposomes or conjugation to synthetic polypeptides such as poly(L-lysine) or by derivatization with carriers such as cholesterol. Coupling the oligonucleotide to specific ligands may enable them to target specific cell populations. One possibility is to encapsulate the oligonucleotide into antibody-targeted liposomes – an antibody incorporated in the liposomes ensures specific attack on those cells expressing the corresponding antigen.

A more novel way of getting the oligonucleotide past the cell membrane and into the cytoplasm uses gene-transfer techniques – in this the oligonucleotide is inserted into the genome of a non-toxic but still invasive virus. Unfortunately, this method rules out chemically modified analogues because they cannot be recognized by the natural cell machinery on which the technique depends.

(4) Specificity and affinity

If the nucleotide sequence of the target molecule is known, then it is possible to write down the chemical formula of the inhibitor corresponding to the base sequence of the ASO, a procedure which amounts to rational drug design. Virtually any region of the RNA can in principle be targeted by the antisense oligonucleotide. However, the ASO must be complementary to exposed regions in their target RNAs and many sites may not be accessible for Crick–Watson base-pairing because of RNA secondary structure or protein binding. Accessibility of sites cannot be predicted and thus an effective antisense molecule still has at present to be found empirically by screening a large number of candidates, perhaps 20–50 molecules, and this is a time-consuming process.

Interaction between the oligonucleotide and its target must be specific. This specificity is determined by the defined base sequence while affinity results from the Crick–Watson hybridization interactions and base stacking in the double helix that forms. Statistically, the base sequence of a 17-mer occurs just once in the human genome and an oligonucleotide of this length hybridizes well with its complementary target mRNA. Modification of an oligonucleotide has an adverse effect on specificity and binding affinity but sequences with 15–20 bases offer the opportunity of extremely selective intervention in gene expression. However, RNase H probably does not require long hybrid regions as substrates and 10 base pairs may be sufficient to lead to cleavage. An ASO of 20 bases also contains 11, 10-nucleotide sequences (10-mers) and a 10-mer will occur in many RNAs. Thus, it is probably not possible to obtain cleavage from an intended target RNA without causing at least partial destruction of many non-target RNAs. However, if a 10-mer complementary to an ASO occurs in an accessible site in a target RNA and in an inaccessible form in a non-target RNA, then the target will be preferentially destroyed. The challenge is to identify anti sense molecules that are complementary to vulnerable sites in target RNAs. This is hard to do.

Whereas the requirements for specificity and binding affinity may be satisfactorily met by the unmodified oligonucleotides, adequate stability to nucleases and sufficient passage through membranes can be achieved only by modification of the oligonucleotides. Phosphorothioates are the most active of the first-generation analogues and have displayed *in vitro* activity against a large number of targets of therapeutic relevance. However, phosphorothioate oligonucleotides have their limitations; the modified links can cause:

(1) Lower (relative to natural phosphodiester oligomers) binding affinity to single-stranded RNA and especially to double-stranded DNA targets: In one of the monophosphorothioate diastereoisomers, the duplex is destabilized because the negatively charged sulphur is directed toward the helix core where it encounters steric and electrostatic repulsions.

(2) Non-sequence-specific-inhibitions: Because of their polyanionic nature, the phosphorothioate oligonucleotides are able to bind to cellular proteins, thereby producing biologically significant non-antisense effects.

(3) Potential toxicity: Digestion by nucleases might lead to mononucleotide phosphorothioates which could be reincorporated into cellular DNA and cause mutations.

Despite these shortcomings, phosphorothioates have displayed adequate pharmacodynamic, pharmacokinetic and toxicological properties in pre-clinical studies to be evaluated in humans. While phosphorothioates are promising as the first generation of ASOs, the search for better analogues continues. Among second-generation compounds are mixed backbone oligomers. These are 'chimeric' molecules containing, for example, a core of phosphorothioate DNA and flanking sequences that are also modified but typically with neutral phosphate analogues. The hope is that the phosphorothioate linkages will preserve RNase H recruitment but that the lower charge density from the neutral analogues will result in increased cellular penetration and nuclease stability and decreased non-specific binding to proteins. An alternative approach which has been effective in preventing enzymatic degradation of ASOs in cell culture is to covalently link the two ends of the molecules together or to ligate two identical molecules.

When an antisense molecule causes a biological effect, it can be difficult to determine whether the change occurred because the reagent interacted specifically with its target RNA or because some non-antisense reaction involving other nucleic acids or proteins was set in motion. Non-antisense effects are not necessarily bad and include potentially useful responses such as stimulation of B-cell proliferation and the inhibition of viral entry into cells. However, it is the unpredictability of non-specific effects which has made it hard to produce drugs that act primarily through true antisense mechanisms.

5.3.3 Clinical studies

The use of antisense molecules to modify gene expression *in vivo* is variable in its efficacy and reliability. However, preliminary results of several clinical studies has demonstrated the safety and to some extent the efficacy of ASOs in patients with malignant diseases. Take, for example, the phosphorothioate antisense agent for the c-*myb* gene, an oncogene implicated in leukaemia. The target protein starts with Met-Ala-Arg-Arg-Pro-Arg-His-Ser-Ile..., and the corresponding gene sequence is $^{5'}$ ATG GCC CGA AGA CCC CGG CAC AGC ATA $^{3'}$. The aim of the phosphorothioate antisense agent, $^{5'}$ TAT GCT GTG CCG GGG TCT TCG GGC $^{3'}$ (**5.14**) is to delete chronic myelogeous leukaemic cells from bone marrow samples to be used for autologous bone marrow transplantation. In bone marrow purging protocols, the marrow is cleansed of leukaemic cells then reinfused into the patient. An ASO targeted to the oncogene should be more efficient than the presently used purging agents in achieving selective elimination of leukaemic cells from the marrow. In phase 1 clinical trials, the bone marrow from 15 leukaemic patients was removed, purged with **5.14** and then reintroduced into the body. Minimal side effects were reported and remission rates were increased by 50%.

Forrivirsen (Vitravene, ISIS 2922), $^{5'}$GTG TTT GCT CTT CTT CTT GCG$^{3'}$, is a 21 nucleotide phosphorothioate that is complementary to sequences in the major intermediate early region of human cytomegalovirus (HCMV) mRNA. This compound is the first phosphorothioate antisense agent to be approved for clinical use. It can be used in the treatment of HCMV retinitis, an opportunistic eye infection that occurs in nearly 30% of AIDS patients.

5.4   Ribozymes

In the antisense DNA approach, the DNA–RNA hybrid may be cleaved by RNase H. This process requires a trimolecular mechanism involving sequential or simultaneous association of the antisense molecule, the target RNA and the RNase in order to achieve cleavage. Ribozymes are RNA enzymes that, as part of the mRNA structure, catalyse the self-splicing of the transcript. Ribozymes therefore do not require a separate enzymatic component for cleavage of the target. Originally discovered by Cech and co-workers in 1981 to play a part in the production of protein from DNA in a protozoan, ribozymes have subsequently been found to be a widespread phenomenon. Of particular interest for therapeutic use are the small pathogenic RNAs that occur in plants. These ribozymes have sequences which fold into distinctive secondary structures termed as 'hammerhead' and 'hairpin'. The hammerhead is the simpler of the two ribozymes and is a roughly γ-shaped motif, which consists of three base-paired stems surrounding a single-stranded central region (Figure 5.4). Stems I and II, the flanking arms of the motif, are not conserved, although they must be correctly base-paired, while some of the nucleotides in the central core are conserved in all hammerheads. Most of the conserved bases cannot form Crick–Watson base-pairs but instead form more complex structures which mediate RNA folding and catalysis. Substitution of any of the conserved bases with naturally occurring bases or with artificial alterations of their functional groups results in diminished catalytic activity.

The hammerhead is a metalloenzyme and requires a divalent metal cation such as Mg^{2+} to mediate catalytic activity. A proposed cleavage reaction is the internal transesterification reaction (Scheme 5.11) which yields one fragment terminating in a 2′,3′-cyclic phosphate while the other has a 5′-hydroxyl terminus.

Hammerhead ribozymes can be modified in their binding arms (stems I and II) to be complementary to any target and therein lies their therapeutic

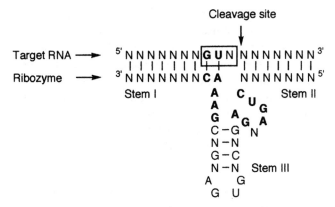

Figure 5.4 Consensus sequence of hammerhead ribozymes, where N represents any base. Residues in bold type are strongly conserved. In most natural hammerhead RNAs, the best triplet (boxed) 5' to the cleavage site is GUC. Alteration to GUA or GUU does not affect cleavage efficiency but cleavage is strongly reduced for a GUG sequence 5' to the cleavage site

potential. The combination of antisense sequences to confer target specificity and catalytic domains to facilitate the cleavage of the target means that ribozymes offer even greater potential effectiveness as inhibitors than antisense DNA. As with the latter, there are many issues that need to be addressed before ribozyme potential can be translated into reality.

(1) Synthesis. Oligoribonucleotides are more elaborate to synthesize than deoxy-oligomers as the presence of the 2'-OH requires a protecting group compatible with those used in the assembly procedure. RNA is sensitive to hydrolysis under alkaline conditions and this further restricts the choice of a suitable protecting group. Assembly of RNA is analogous to that of DNA

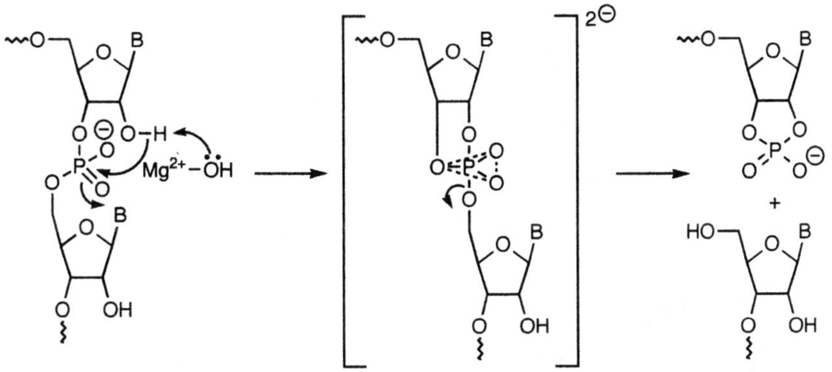

Scheme 5.11 Possible mechanism of RNA cleavage by Mg^{2+} ions. Coordination of the Mg^{2+} to the RNA facilitates abstraction of the 2'-OH proton which initiates nucleophilic attack at the phosphorus. This leads to formation of a pentacoordinate transition state with a trigonal-bipyramidal structure which, in turn, generates products having 2',3'-cyclic phosphate and 5-hydroxyl termini

except that coupling reactions are slower and this necessitates either an increase in coupling time or the use of more powerful activating and coupling reagents. For example, in phosphoramidite assembly with *t*-butyldimethylsilyl (TBDMS) protection for the 2′-OH function (Scheme 5.12), the more active 4-nitrophenyltetrazole is used in place of tetrazole to effect coupling. Pivaloyl chloride is again used as activator in H-phosphonate chemistry. The TBDMS group is removed by $Bu_4N^+F^-$ (TBAF) after chain assembly is complete. Synthetic RNAs are often stored in the 2′-protected form to prevent the possibility of inadvertent degradation by ribonucleases.

(2) Nuclease stability. RNAs are susceptible to degradation by ribonucleases. They can be chemically modified with, e.g. phosphorothioates or with a 2′-OMe group, but the effect of these alterations on the catalytic activity is uncertain.

(3) Cellular uptake and delivery. These are major challenges. Ribozymes can be delivered exogenously to cells as preformed molecules or endogenously by

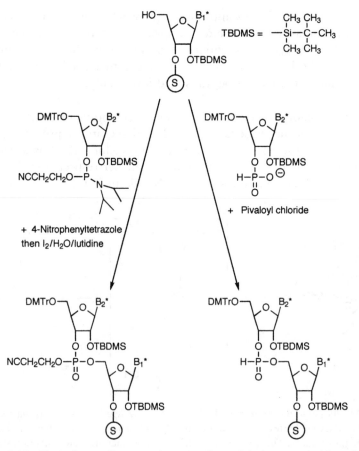

Scheme 5.12 Typical coupling reactions in the solid-phase assembly of RNAs by phosphoramidite and H-phosphonate chemistry (B*, protected base)

introducing their DNA templates by means of a viral vector. Efficient delivery of synthetic RNA into target cells may be achieved by adapting the existing repertoire of techniques developed with antisense DNA – covalent linking of lipophilic groups or incorporation into liposomes. Targeting of specific cells types or organs, colocalization of the ribozyme with the target, and enhancing substrate turnover are other factors which will also ultimately determine the effectiveness of the ribozyme.

(4) Specificity. A recognition sequence of approximately 15 nucleotides, seven in each of the 3′- 5′-arms and one at the core, should confer reasonable uniqueness and therefore therapeutic specificity.

Ribozymes have not been investigated with a view to therapeutic use as intensively as has antisense DNA but the targets are similar. Viral infections where the target gene is foreign are particularly attractive targets, as are cancers where the disease results from mutation of a protooncogene since the new gene product is not formed in the normal adult cell. It remains to be seen whether ribozymes are at least as effective as the quasi-catalytic action of ASOs involving cleavage of the target RNA by endogenous RNase H activity.

Further Reading

Textbook and review articles

- S. Abdulla, Silencing the Code, *Chem. Brit.*, 1997, **33** (July), 30–33.
- S. Agrawal, Antisense Oligonucleotides: Towards Clinical Trials, *Trends Biotechnol.*, 1996, **14**, 376–387.
- S. Agrawal and R. P. Iyer, Modified Oligonucleotides as Therapeutic and Diagnostic Agents, *Curr. Opin. Biotechnol.*, 1995, **6**, 12–19.
- R. M. Blaese, Gene Therapy for Cancer, *Sci. Am.*, 1997, **276** (June), 91–95.
- A. D. Branch, A Good Antisense Molecule is Hard to Find, *Trends Biochem. Sci.*, 1998, **23**, 45–50.
- R. E. Christoffersen and J. J. Marr, Ribozymes as Human Therapeutic Agents, *J. Med. Chem.*, 1995, **38**, 2023–2037.
- S. T. Crooke, Therapeutic Applications of Oligonucleotides, *Annu. Rev. Pharmacol. Toxicol.*, 1992, **32**, 329–376.
- S. T. Crooke and C. F. Bennett, Progress in Antisense Oligonucleotide Therapeutics, *Annu. Rev. Pharmacol. Toxicol*, 1996, **36**, 107–129.
- R. G. Crystal, Transfer of Genes to Humans: Early Lessons and Obstacles to Success, *Science*, 1995, **270**, 404–410.
- F. Eckstein (Ed.), *Oligonucleotides and Analogues: A Practical Approach*, IRL Press, Oxford, UK, 1991.
- D. L. Ennist, Gene Therapy for Lung Disease, *Trends Pharmacol. Sci.*, 1999, **20**, 260–266.

- T. Friedmann, Overcoming the Obstacles To Gene Therapy, *Sci. Am.*, 1997, **276** (June), 80–85.
- U. Galderisi, A. Cascino and A. Giordano, Antisense Oligonucleotides as Therapeutic Agents, *J. Cell. Physiol.*, 1999, **181**, 251–257.
- L. Kværnø and J. Wengel, Antisense Molecules and Furanose Conformations – Is It Really that Simple? *Chem. Commun.*, 2001, 1419–1424.
- S. Maulik and S. D. Patel, Human Gene Therapy, in *Molecular Biotechnology: Therapeutic Applications and Strategies*, Wiley, New York, 1997, pp. 37–67.
- A. D. Miller, Cationic Liposomes for Gene Therapy, *Angew. Chem. Int. Ed. Engl.*, 1998, **37**, 1768–1785.
- P. D. Robbins, H. Tahara and S. C. Ghivizzani, Viral Vectors for Gene Therapy, *Trends Biotechnol.*, 1998, **16**, 35–40.
- A. E. Smith, Viral Vectors in Gene Therapy, *Annu. Rev. Microbiol.*, 1995, **49**, 807–838.
- R. H. Symons, Small Catalytic RNAs, *Annu. Rev. Biochem.*, 1992, **61**, 641–671.
- P. Tolstoshev, Gene Therapy, Concepts, Current Trials and Future Directions, *Annu. Rev. Pharmacol. Toxicol.*, 1993, **32**, 573–596.
- E. Uhlmann and A. Peyman, Antisense Oligonucleoticles: A New Therapeutic Principle, *Chem. Rev.*, 1990, **90**, 543–584.
- J. A. Wagner, A. C. Chao and P. Gardner, Molecular Strategies for Therapy of Cystic Fibrosis, *Annu. Rev. Pharmacol. Toxicol.*, 1995, **35**, 257–276.

Research publications

- R. M. Blaese, K. W. Culver, A. D. Miller, C. S. Carter, T. Fleisher, M. Clerici, G. Shearer, L. Chang, Y. Chiang, P. Tolstoshev, J. J. Greenblatt, S. A. Rosenberg, H. Klein, M. Berger, C. A. Mullen, W. J. Ramsey, L. Muul, R. A. Morgan and W. F. Anderson, T Lymphocyte-Directed Gene Therapy for ADA-SCID: Initial Trial Results After 4 Years, *Science*, 1995, **270**, 475–480.
- C. Bordignon, L. D. Notarangelo, N. Nobili, G. Ferrari, G. Casorati, P. Panina, E. Mazzolari, D. Maggioni, C. Rossi, P. Servida, A. G. Ugazio and F. Mavilio, Gene Therapy in Peripheral Blood Lymphocytes and Bone Marrow for ADA⁻ Immunodeficient Patients, *Science*, 1995, **270**, 470–475.
- M. Cavazzana-Calvo, S. Hacein-Bey, G. de Saint Basile, F. Gross, E. Yvon, P. Nusbaum, F. Selz, C. Hue, S. Certain, J.-L. Casanova, P. Bousso, F. Le Deist and A. Fischer, Gene Therapy of Human Severe Combined Immunodeficiency (SCID)-XI Disease, *Science*, 2000, **288**, 669–672.
- Z. Ram, K. W. Culver, E. M. Oshiro, J. J. Viola, H. L. DeVroom, E. Otto, Z. Long, Y. Chiang, G. J. McGarrity, L. M. Muul, D. Katz, R. M. Blaese and E. H. Oldfield, Therapy of Malignant Brain Tumors by Intratumoral Implantation of Retroviral Vector-Producing Cells, *Nat. Med.*, 1997, **3**, 1354–1361.

6
Oligosaccharides

6.1 Overview

Many pharmaceutical products contain carbohydrate or modified carbohydrate components but there are few examples of pure carbohydrate therapeutics. Heparin (**6.1**) is one of the few examples of a drug which is composed solely of sugar units. It is an anticoagulant which is used clinically to treat thromboembolic disorders (see Section 6.3 below). Carbohydrate-containing pharmaceuticals are much more common than pure polysaccharide therapeutics and have a wide range of biological activities. Streptomycin (**6.2**) is an aminoglycoside antibiotic which is principally used as an antituberculotic agent. The anthracyclines doxorubicin (**6.3a**) and daunorubicin (**6.3b**) are antibiotics with antitumour properties and the ene–diyne calicheamicin γ_1 (**6.4**) is a promising new anticancer agent. The cardioactive glycosides represent another class of medicinal agents possessing carbohydrate components that contribute to their therapeutic efficacy. Digoxin (**6.5a**), isolated

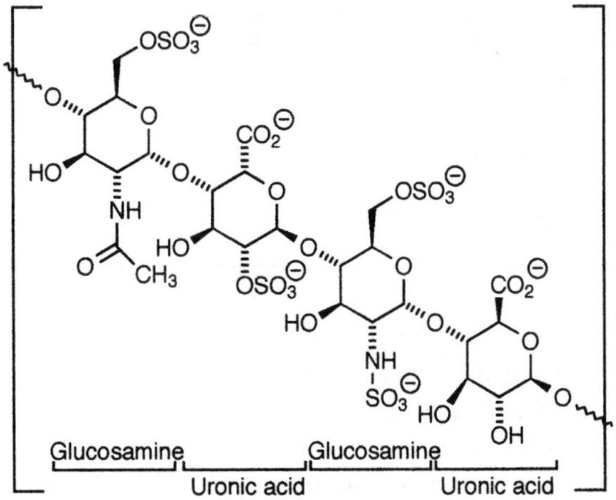

(**6.1**) Heparin

169

(6.2) Streptomycin

(6.3) (a) Doxorubicin R = OH; (b) daunorubicin R = H

(6.4) Calicheamicin γ_1

(6.5) Cardiac glycosides. (a) Digoxin: R = α-OH, R^1 = OH; (b) digitoxin: R = β-OH, R^1 = H

from *Digitalis lanata*, and digitoxin (**6.5b**), isolated from *D. purpurea*, are used in the treatment of heart arrhythmias. Inhibition of gene expression can be mediated by synthetic antisense oligonucleotides, compounds which can also be classed as carbohydrate-containing therapeutics.

Carbohydrates can also be covalently attached to proteins to form glyco-proteins (see Section 6.4 below) or to different types of lipids to form glycolipids. Glycoproteins are found as soluble compounds, linked to cell membranes, inside cells or in extracellular fluids. Many of the proteins discussed in Chapters 2–4 (i.e. clotting factors, hormones, enzymes and immunoglobulins) are glycoproteins. The carbohydrate groups of glyco-proteins confer important physical properties such as stabilization of confor-mation, protease resistance, charge and water-binding capacity. Oligosacchar-ide chains of cell membrane glycoproteins and glycolipids play important roles in intercellular recognition processes, acting as receptors for proteins, enzymes and viruses and serving as determinants of immunological specificity. The blood group antigens are some of the best characterized examples of carbo-hydrate cell-surface antigens (see Section 6.4.1 below).

Alterations to the oligosaccharide pattern of glycoconjugates on the outer surface of cell membranes is implicated in many pathological processes, including malignancy. The identification of stable polysaccharide antigens on bacterial cell surfaces has led to the development of several carbohydrate-based anti-bacterial vaccines. Target infections are *Haemophilus influenzae* and *Neisseria meningitidis*, major causes of meningitis in young children, and *Streptococcus pneumoniae* (see Section 6.5 below). These vaccines stimulate an immune response, including the production of antibodies which provide protection from subsequent infection. Similarly, specific carbohydrate antigens overexpressed in cancer cells have become targets for the development of cancer vaccine therapies (see Section 6.6 below).

6.2   Oligosaccharide Synthesis

Pharmaceutical preparations of heparin and the other compounds mentioned above are obtained from natural sources rather than synthesized. Compared with the synthesis of peptides and oligonucleotides, the synthesis of oligosac-charides and complex carbohydrate-containing molecules is far more difficult. If two identical amino acids or nucleotides are joined together, only one dipeptide or dinucleotide is obtained, whereas when two identical monosac-charides such as glucose are linked, 11 different disaccharides are possible, and that considers only the most frequently occurring pyranose form. Furthermore, carbohydrate polymers can also be branched and this structural feature is not found in the other basic biopolymers. Synthesis problems are derived from the multifunctional nature of the compounds and the stereo-control required during glycoside coupling.

The synthesis of oligosaccharides involves the coupling of two saccharide units in a process whereby one of them (6.6) functions as a glycosyl donor and the other (6.7) as a glycosyl acceptor (Scheme 6.1). The key reactive intermediate is the cyclic oxocarbenium ion 6.8 and many glycosyl donors have been employed to form this species.

Hydroxide is not a good leaving group and has to be replaced by other groups to activate the electrophilic partner. The classic anomeric leaving

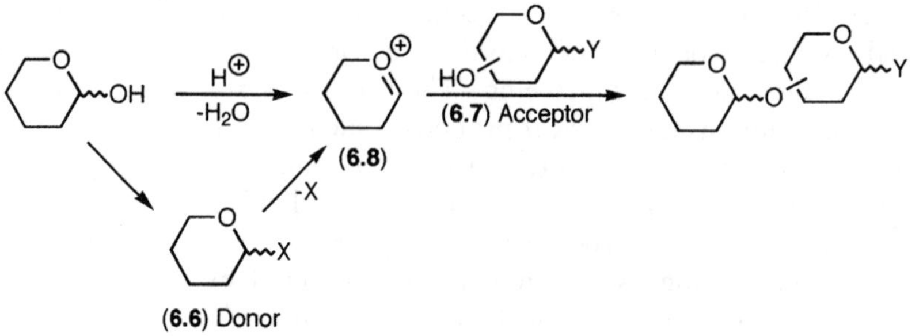

Scheme 6.1   Glycoside bond-forming reaction

group of the glycosyl donor is halide. Glycosyl bromides or chlorides are converted into the oxocarbenium ion by the action of heavy metal ions, typically silver or mercury, and couple with the hydroxy component in a method of glycosidation known as the Koenigs–Knorr reaction (Scheme 6.2(a)). Depending on the relation between the groups at C-1 and C-2, the coupling step can furnish linkages of the 1,2-*cis* or 1,2-*trans* type in the glycoside produced. Synthesis of 1,2-*trans* glycosides is generally straightforward and makes use of 2-*O*-acyl substituents in neighbouring group participation (Scheme 6.2(b)). Formation of linkages of the 1,2-*cis* type is more exacting and involves *in situ* conversion of an α-halide to the more reactive β-halide and subsequent glycosidation with inversion of configuration (Scheme 6.2(c)). Carbohydrates possess multiple hydroxyl groups and regioselective glycosylation of acceptor sugars can be dealt with by selective protection strategies. Hydroxyl groups are most often protected by formation of ethers (frequently benzyl) or esters (with acetyl and benzoyl being the most common). Diols can also be simultaneously protected by conversion to cyclic acetals, the most common of which are the benzylidene (for 1,3-diols) and isopropylidene (for 1,2-diols) derivatives.

Koenigs–Knorr glycosidations are widely used but the reaction is not without its disadvantages – glycosyl halides need to be prepared and used directly and they are sensitive to hydrolysis, and the choice of catalyst can have a significant bearing on the stereoselectivity of syntheses utilizing *in situ* anomerization.

The preparation of a heparin pentasaccharide in the mid 1980s using the classical silver-catalysed glycosidation reaction in a convergent approach typifies oligosaccharide syntheses at that time (Scheme 6.3). Disaccharides corresponding to rings EF and GH of the pentasaccharide were constructed first, then coupled together to form a tetrasaccharide and finally ring D was added. The heparin pentasaccharide synthesized was a tri-*N*-sulphated structural variant of the sequence that binds to antithrombin III since replacement of the *N*-acetyl group of fragment D of heparin by a sulphamido group occurs in beef-lung heparin and homogenous sulphation of the three amino functions simplified the synthesis strategy. Acetyl protection was selected for the

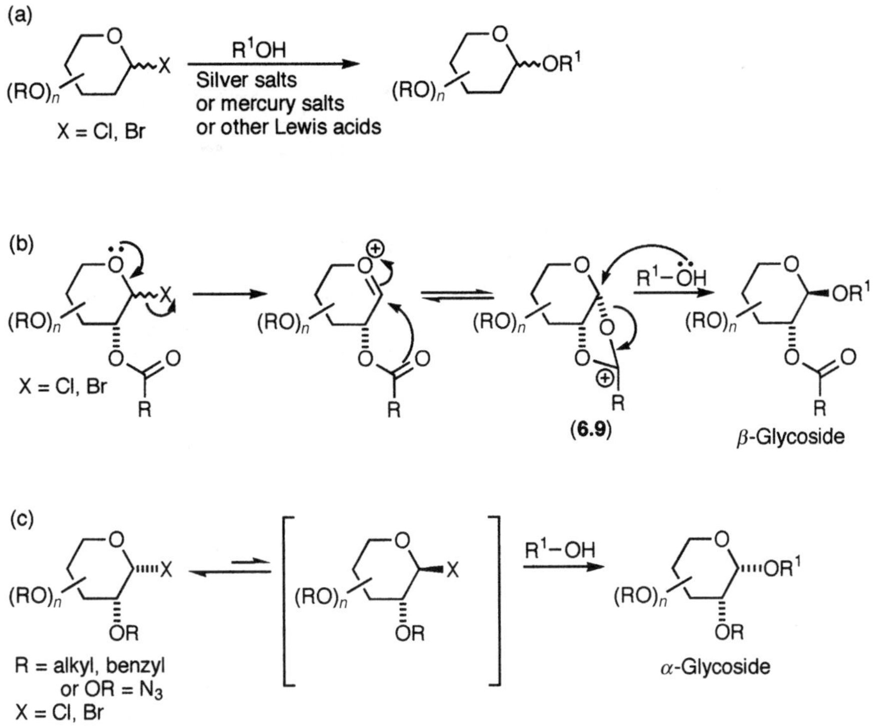

(a)

(b)

(6.9)   β-Glycoside

(c)

R = alkyl, benzyl
    or OR = N₃
X = Cl, Br

α-Glycoside

Scheme 6.2 The Koenigs–Knorr glycosidation reaction. (a) Classically, the anomeric halides are activated by silver or mercury salts but modern variants of the conditions include the use of other Lewis acids. The anomeric stereochemistry is controlled by the nature of the C-2 substituent. (b) Formation of 1,2-*trans* glycosides. When the C-2 position is occupied by a group such as an ester which is capable of neighbouring group participation, the halide, irrespective of whether the α- or β forms are used, reacts, via the oxocarbenium ion, to give the stable cyclic ion 6.9. Nucleophilic ring opening with the hydroxylic component then leads (for steric reasons) to glycosides of 1,2-*trans* type (often β-glycosides). (c) Formation of 1,2-*cis* glycosides. When the C-2 oxygen is protected with an alkyl or benzyl group or if the C-2 position is occupied by another non-participating group such as azido (N₃), the anomeric effect dominates and the α-anomer is preferentially formed. The anomeric effect is particularly strong for halides and in solution the stable and isolable α-halides are in equilibrium, via various intermediates, with the less stable but more reactive β-form. While there is a higher proportion of α-halide to β-halide at equilibrium, the latter is more reactive towards the hydroxylic component so that formation of a 1,2-*cis* link (often α-glycosides) is faster than that of the *trans* glycoside

hydroxy groups which would ultimately be sulphated with the other hydroxy groups permanently protected as benzyl ethers. The amino groups of the glucosamine units were protected as either the benzyloxycarbonyl derivative or masked as an azide. In the Koenigs–Knorr glycosidation reactions, formation of the *trans*-glycosidic linkage between E and F was promoted by silver carbonate but the more reactive silver triflate was required to introduce the

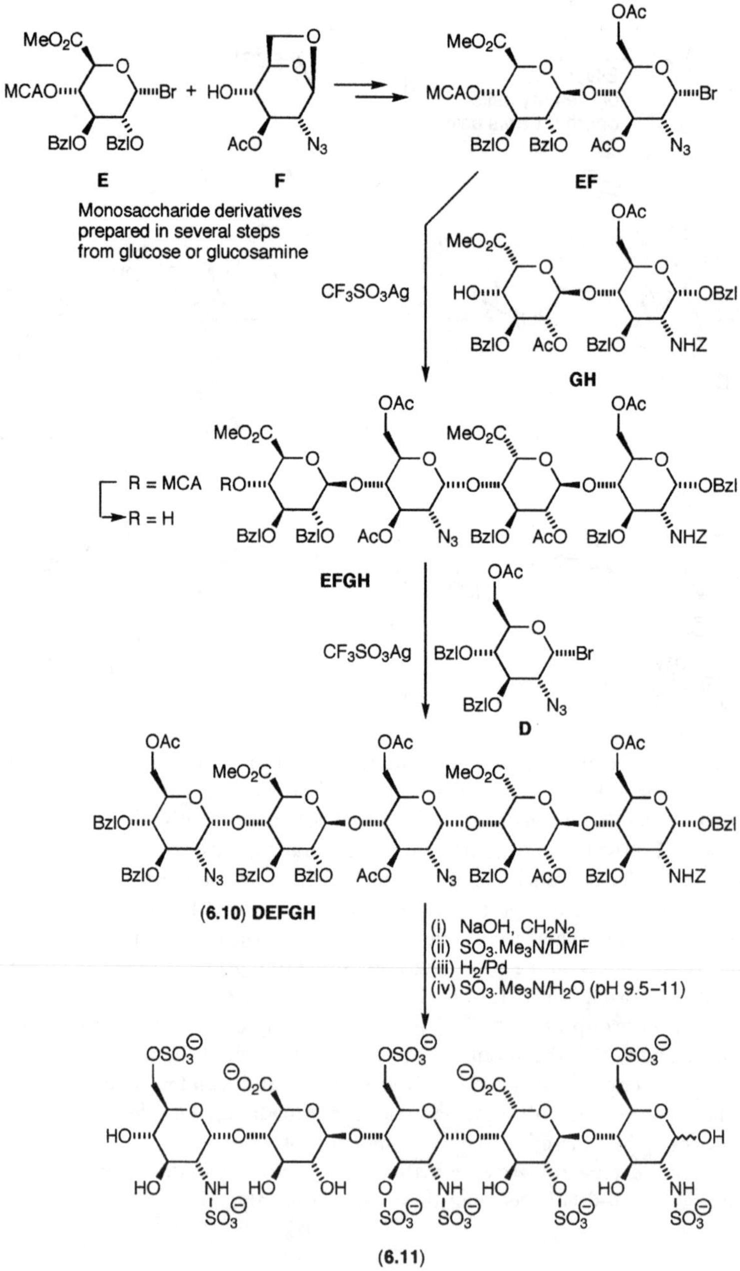
E
F
Monosaccharide derivatives prepared in several steps from glucose or glucosamine
EF
CF3SO3Ag
GH
R = MCA
R = H
EFGH
CF3SO3Ag
D
(6.10) DEFGH
(i) NaOH, CH2N2
(ii) SO3.Me3N/DMF
(iii) H2/Pd
(iv) SO3.Me3N/H2O (pH 9.5–11)
(6.11)

Scheme 6.3 Synthesis of heparin pentasaccharide (MCA, monochloroacetyl ($-COCH_2Cl$))

EF–GH and D–EFGH *cis* linkages. The fully protected pentasaccharide (**6.10**) was converted into the biologically active pentasaccharide (**6.11**) by the following sequence of reactions: (i) *O*-deacetylation, (ii) sulphation of the free hydroxyl groups, (iii) hydrogenolysis to generate the amino groups of the glucosamine units and to liberate hydroxyl groups, and (iv) *N*-sulphation of amino groups. The overall yield of **6.11** starting from glucose was less than 0.1%.

Over the last 20 years there has been vigorous activity in the field of oligosaccharide synthesis methodology, much of it focused on the glycosidation reaction. New glycosyl donors have been developed which are both easy to form and which have extended shelf-lives, which react with glycosyl acceptors under mild conditions and which offer a greater degree of stereocontrol, particularly in α-glycosidation. Trichloroacetimidates (**6.12**), thioglycosides (**6.13**) (particularly phenylthio- and ethylthioglycosides), *n*-pentenyl glycosides (**6.14**) and glycals (**6.15**) all satisfy these requirements to a lesser or greater degree and the synthesis of oligosaccharides comprising up to five sugars is now considered routine.

(**6.12**) Trichloroacetimidate

(**6.13**) Thioglycoside

(**6.14**) *n*-Pentenyl glycoside

(**6.15**) Glycals

In the glycal method, the electron-rich olefin of the enol ether is treated with an electrophilic reagent (E^+) to form a 3-membered 'onium ion' species which is then glycosidated by the acceptor sugar (Scheme 6.4(a)). The electrophile can then be removed reductively to form the 2-deoxyglycoside. The Danishefsky variation of the glycal method (Scheme 6.4(b)) uses the electrophile dimethyldioxirane which gives access to both 2-oxo and 2-deoxyglycosides. Glycals have been extensively used by Danishefsky and co-workers in the synthesis of many oligosaccharides including tumour-associated antigens (see Section 6.6 below).

Enzymic methods of glycosidation have also been successfully used to construct oligosaccharides. Activation of the anomeric position of the monosaccharide is achieved by esterification by a phosphate group of a nucleotide

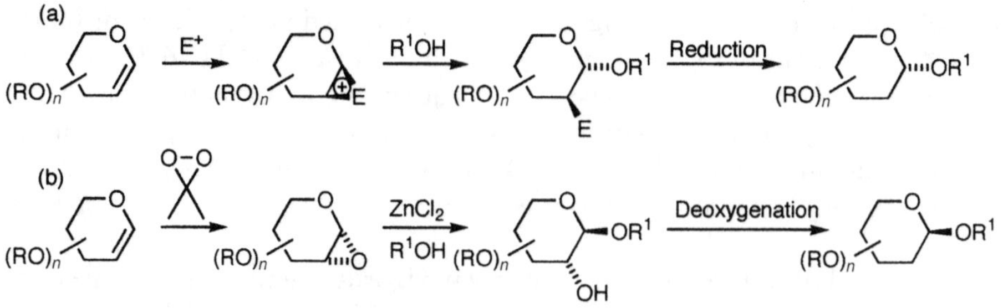

(a)

(b)

Scheme 6.4 The glycal-based glycosidation reaction. (a) Original method, $E^+ = I(\text{collidine})_2 ClO_4$, N-iodosuccinimide, or PhSeCl. In most cases, α-glycosides are formed due to preferential top-face attack by the electrophile on the glycal. (b) Danishefsky method. 3,3-Dimethyldioxirane has a preference for α-attack on the glycal and so β-glycosides are formed due to inversion of configuration at the glycosidation step

and the reaction is catalysed by a glycosyl transferase (Scheme 6.5). Glycosyl transferases are substrate-specific which allows for high chemoselectivity in the glycosidation reaction, and stereospecific which allows for formation of either the α- or β-anomer exclusively. Such selectivity means that oligosaccharides can be synthesized from unprotected sugars. While a number of transferases have been cloned and can be obtained from cultured cells, efficient syntheses can be achieved with partially purified enzymic preparations extracted from mammalian organs. Sugar nucleotides are prohibitively expensive for use as stoichiometric reagents. Non-stoichiometric amounts can be used and then the nucleotide phosphate byproduct (**6.16**) from the glycosidation reaction can be enzymically combined with (inexpensive) sugars to generate nucleotide sugar phosphates *in situ*.

Oligosaccharide synthesis has been adapted to solid-supported approaches but it is nowhere near as efficient as solid-phase peptide and oligonucleotide synthesis. Each glycosidation must ideally proceed in very high yield with complete stereocontrol and this is still difficult to achieve. There are two strategies wherein either the glycosyl acceptor or glycosyl donor is immobilized on the solid support and glycosidation achieved by addition of an excess

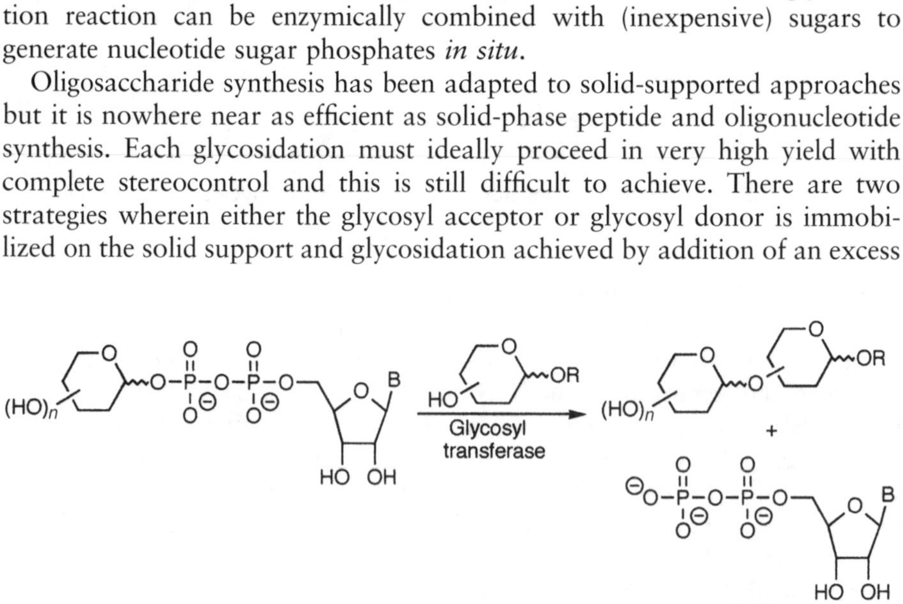

(**6.16**)

Scheme 6.5 Enzyme-catalysed glycoside bond-forming reaction of glycosyl donor or acceptor, respectively, together with a promoter. The process is reiterated to assemble the desired sequence and this is then followed by cleavage from the polymeric support and purification to give the oligosaccharide. Danishefsky's glycal/glycal epoxide methodology has worked particularly well on the solid phase (Scheme 6.6). The glycosyl donor (6.17) is polymer-bound and is activated by epoxidation with dimethyldioxirane. Treatment of the epoxide with the glycosyl acceptor *as a glycal* in the presence of zinc chloride gives glycoside 6.18 with the glycosyl donating function already in place for the next coupling event. The process can be terminated by adding a conventional sugar rather than a glycal as acceptor. Furthermore, opening of the epoxide during glycosidation exposes a C-2 hydroxyl group which can, in turn, serve as a glycosyl acceptor to form branched oligosac-

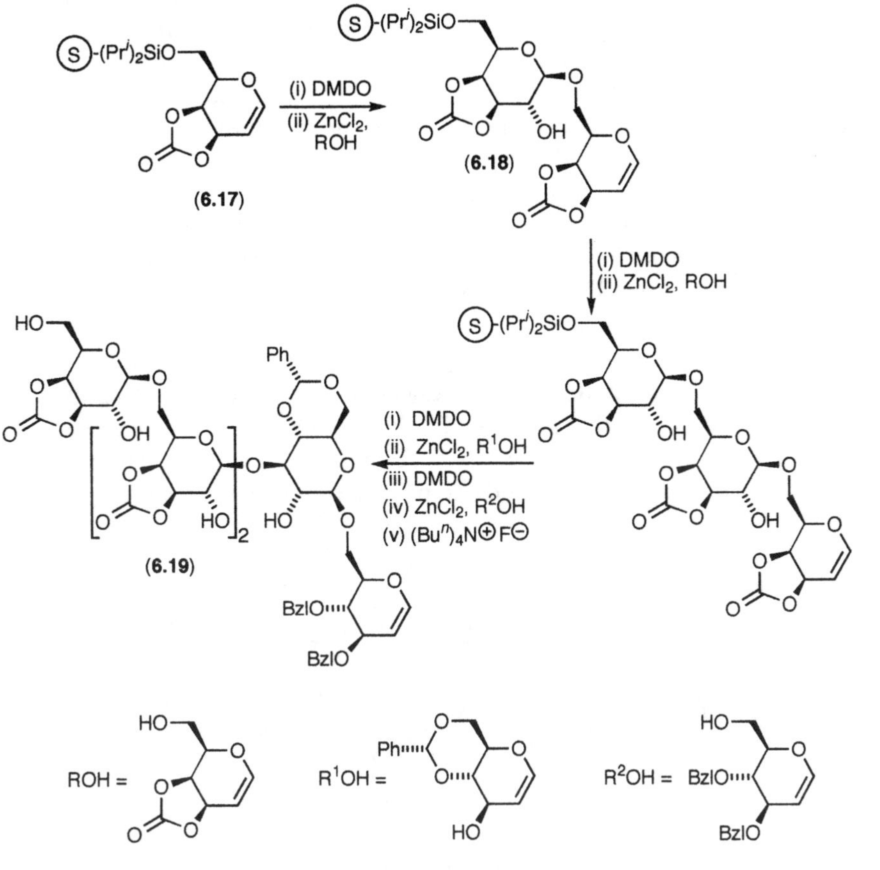

Scheme 6.6 Solid-phase synthesis of a pentasaccharide by an iterative glycal glycosidation reaction. The first glycal is attached to a polystyrene resin by using a silyl ether linker. The oligosaccharide is built up by using a two-step process of epoxidation using dimethyldioxirane (DMDO), followed by glycosidation with the appropriate glycosyl acceptor. The pentasaccharide 6.19 was obtained in 39% overall yield from the polymer-linked glycal 6.17

charides. Secondary hydroxyl groups are less reactive than primary ones and so the presence of such an unprotected group on sugar fragments is not a problem during linear syntheses. The solid-phase synthesis of the pentasaccharide **6.19**, outlined in Scheme 6.6, illustrates how the rapid synthesis of complex oligosaccharides is within reach.

6.3   Heparin

The polysaccharide heparin can be considered the body's natural anticoagulant as it is involved in maintaining the fluidity of the blood. Because of its anticoagulant properties, heparin has been used clinically for the treatment and prophylaxis of thrombosis. Heparin is a heterogeneous mixture of highly sulphated glycosaminoglycan polymers with molecular weights ranging from 6000 to 30 000 Da. Individual chains are built up of long linear sequences of predominantly alternating 1,4-linked uronic acid (D-glucuronic or L-iduronic) and D-glucosamine or N-acetyl-D-glucosamine residues that carry sulphate substituents at various positions. Heparin exerts its anticoagulant effects by binding to antithrombin III (AT III), a serine protease inhibitor. This naturally occurring plasma glycoprotein forms tight complexes with, and thereby inactivates, several serine protease blood coagulation factors, including factors Xa and IIa (thrombin). The rate of protease inactivation is increased by several orders of magnitude when AT III is bound to a heparin molecule. The 'catalyst' is a pentasaccharide domain (**6.20**) of heparin. Binding of this pentasaccharide region to AT III induces a conformational change in the protein which causes the protease binding loop to become exposed. This activated form of AT III then binds with the aforementioned factors (Figure 6.1) to give a stable complex and the pentasaccharide is released.

(**6.20**) Pentasaccharide domain of heparin

Thus heparin, released naturally in *vivo* or administered therapeutically, inhibits the blood coagulation cascade. Pharmaceutical preparations of heparin have been used since 1937 and are obtained from the lungs and

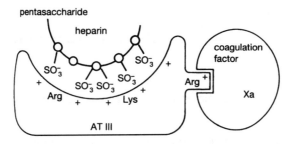

Figure 6.1 Heparin-mediated activation of antithrombin III and reaction with factor Xa. Reproduced from van Boeckel, C. A. A., and Petitou, M., *Angewandte Chemie, International Edition—English*, 32, 1671–1690, 1993, by permission of the authors.

intestinal tissue of slaughtered cattle and pigs. A common side-effect of heparin use is haemorrhage. The anticoagulant activity of heparin is related to the molecular weight of the polysaccharide molecules and the haemorrhagic problems can be reduced by using low-molecular-weight heparins (LMWHs). The latter contain the unique pentasaccharide required for specific binding with AT III but in lower proportions than those found in the standard heparin preparations. LMWHs have molecular weights in the 3000 to 6000 range and are obtained by chemical (hydrolysis with nitrous acid) or enzymatic (heparinase) cleavage of standard heparin chains.

6.4 Glycoproteins

The carbohydrate is attached to the protein component of the glycoproteins through a β-N-glycosidic bond to the amide group of asparagine or by an α-O-glycosidic linkage with the side-chain oxygen atom of serine or threonine residues (Figure 6.2). Oligosaccharide units attached to proteins by N-glycosidic linkages contain, as a rule, a common pentasaccharide core (**6.21**) attached to the asparagine in a Asn X–Ser(Thr) sequence, but not all

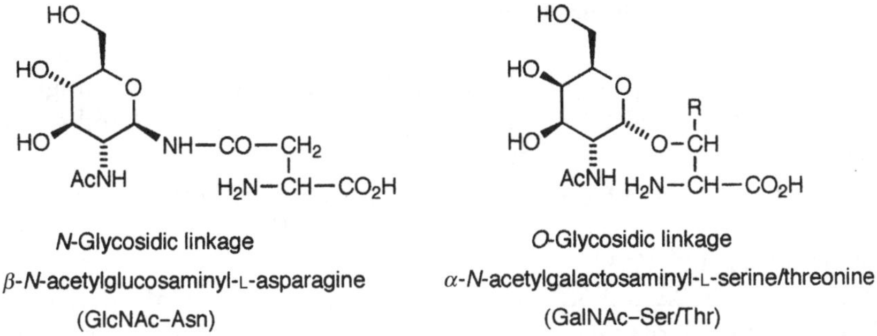

N-Glycosidic linkage

β-*N*-acetylglucosaminyl-L-asparagine

(GlcNAc–Asn)

O-Glycosidic linkage

α-*N*-acetylgalactosaminyl-L-serine/threonine

(GalNAc–Ser/Thr)

Figure 6.2 Carbohydrate–amino acid linkages in glycoproteins

Man-α-(1-6) $\diagdown$

            Man-β-(1-4)-GlcNAc-β-(1-4)-GlcNAc-β-(1-Asn)

Man-α-(1-3) $\diagup$

(6.21) Common core sequence of *N*-glycoproteins

sequences of this type are necessarily glycosidated. In contrast to *N*-glycosides there are few, if any, generalizations which can be made about *O*-glycosides. An α-D-GalNAc unit is frequently linked to the hydroxyl group of serines or threonines but there is no known consensus sequence specifying *O*-glycosylation sites.

The carbohydrate content of glycoproteins varies but may comprise up to 60% of the mass of the molecule. For example, 31% of the hormone chorionic gonadotrophin (molecular weight 38 000) is carbohydrate while immunoglobulins (150 000 Da) and interferon (26 000 Da) have a carbohydrate content of 20% and 10%, respectively. Oligosaccharides are not primary gene products but are synthesized by glycosyltransferases, enzymes that transfer sugars from their nucleotides, for example, uridine diphospho-galactose (UDP-Gal) (see Section 6.4.1 below). This type of synthesis is less accurate than protein synthesis and results in microheterogeneity in the carbohydrate moiety of glycoproteins. The glycosylation pattern of a protein can affect a protein-based drug's potency. Lack of glycosylation by microorganisms has mandated the use of mammalian cells for the production of some rDNA-derived proteins instead of bacterial fermentations. Many protein-based therapeutics either fail to function or cause immune reactions when not properly glycosylated. Different glycoforms of tPA, which is used to dissolve blood clots following heart attacks and strokes, also have differences in their biological activities. Type 1 tPA, containing three N-linked oligosaccharide chains, is about 30% less effective in plasminogen activation than type 2 tPA, which has two N-linked oligosaccharide chains.

6.4.1   Blood group oligosaccharides

An individual's blood group is determined by the carbohydrate antigens on the surface of red blood cells. The difference between A, B, AB and O blood groups lies in the structure of the outer, and immunodominant, sugar of the oligosaccharide chains of glycoproteins or glycolipids. Different individuals carry either the A antigen (**6.22**), the B antigen (**6.23**), both (group AB), or the H antigen (**6.24**) (group O).

Trisaccharides **6.22** and **6.23** are glycosidation products of the disaccharide **6.24** at position 3 of galactose by an *N*-acetylgalactosamine unit and a galactose unit, respectively. A, B and O genes are not responsible for direct production of the blood group determinant but rather they are responsible for

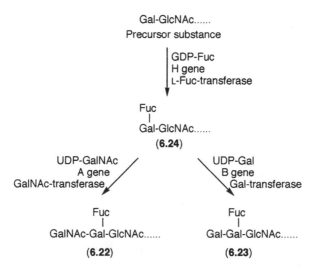

(6.22) A antigen   (6.23) B antigen   (6.24) H antigen

the formation of enzymes that attach the outer sugar to existing oligosaccharide chains (Scheme 6.7). Individuals who carry the A gene possess an N-acetylgalactosaminyltransferase that attaches an N-acetylgalactosamine residue to a precursor with blood group type O specificity; people who carry the B gene possess a galactosyltransferase that attaches a galactose residue to the same O-specific precursor. Individuals who carry both genes posses both transferases, and thus are of AB type. When both genes are absent, the individuals are of blood group O.

In blood group A individuals, the B molecule is recognized as a foreign substance and gives rise to the appearance of anti-B antibodies. Likewise, the A molecule gives rise to anti-A antibodies in individuals with blood group B while anti-A and anti-B antibodies are found in people with blood group O.

Gal-GlcNAc......
Precursor substance

GDP-Fuc
H gene
L-Fuc-transferase

Fuc
|
Gal-GlcNAc......
(6.24)

UDP-GalNAc
A gene
GalNAc-transferase

UDP-Gal
B gene
Gal-transferase

Fuc
|
GalNAc-Gal-GlcNAc......
(6.22)

Fuc
|
Gal-Gal-GlcNAc......
(6.23)

Scheme 6.7 Biosynthesis of the A, B and H(O) blood group determinants. The determinants of blood type are secondary gene products. The primary gene products are the sugar transferases, and it is these enzymes, i.e. N-acetylgalactosaminyltransferase, L-fucosyltransferase and galactosyltransferase, working in concert that determine which structures are formed (UDP, uridine diphosphate; GDP, guanosine diphosphate)

Normally it is therefore only possible to transfuse blood between individuals of the same blood group. However, individuals with AB blood can accept blood of any ABO type, because they lack antibodies for A and B antigens, while those with O blood are universal blood donors because they do not have A and B antigens.

6.5   Polysaccharide Bacterial Vaccines

Bacteria express carbohydrate antigens as components of glycolipids on the cell surface and/or as an outer protective polysaccharide capsule. Both types consist of repeating oligosaccharide units, ranging in size from di- to hexa-saccharides. The antigenic cellular material can be removed from the bacterial cells and forms the basis of antibacterial vaccines. Most bacterial strains express different capsular and/or glycolipid structures which result in the generation of different serotypes. Multivalent vaccines are consequently required for full protection because their antigens are not cross-immunizing; the immune system must mount a separate immune response to each strain. The commonly used vaccine against *Streptococcus pneumoniae* contains purified polysaccharide capsular antigens from 23 types of the organism. Meningococcal vaccines used for active immunization against *Neisseria meningitidis* infections which include meningitis and septicaemia are generally bivalent or tetravalent.

Vaccines prepared from pure polysaccharides are poor immunogens, especially in children less than two years of age. Antibody response is poor and immunological memory is not induced and this makes effective vaccination of young children difficult. The immunogenicity of pure polysaccharides can be enhanced by covalently linking them to a protein carrier to form a conjugate vaccine. Haemophilus influenzae (Hib) conjugate vaccines are used for active immunization against *Haemophilus influenzae* type b infections which causes the most common type of bacterial meningitis and is a major cause of systemic disease in young children. The Hib conjugate vaccine is now part of childhood immunization schedules and is prepared by linking the purified capsular polysaccharide of *H. influenzae* type b to a protein carrier.

Although currently used vaccines are bacterial in origin, there is the possibility that synthetic material could substitute for bacterial isolates in the future.

6.6   Approaches to Carbohydrate-Based Cancer Vaccines

Aberrant cell-surface glycosylation is often strongly associated with the onset of cancer and some tumour-associated carbohydrate antigens have become targets for the development of cancer vaccine therapies. Among these are globo H, an antigen commonly found on breast cancer cells, Lewis[y] (Le[y]) a blood group determinant that is overexpressed on the majority of carcinomas and mucin-core structures, e.g. the Thompson–Friedenreich (TF) disaccharide and the Tn monosaccharide. The idea is that patients immunized with synthetic carbohydrate vaccines would produce antibodies reactive with cancer cells and that the production of such antibodies would mitigate against tumour spread, so enabling a more favourable prognosis. Experimental antitumour vaccines developed by Danishefsky and coworkers have generally contained synthetic antigens (6.25–6.28) conjugated to a carrier protein, usually keyhole limpet haemocyanin (KLH), combined with an adjuvant. Small molecules such as short oligosaccharides often are not able to generate an immune response on their own but they can be made immunogenic by coupling to a carrier protein. Conjugation can be direct, by coupling the native reducing ends of carbohydrates (or generated aldehydes) to amine groups on

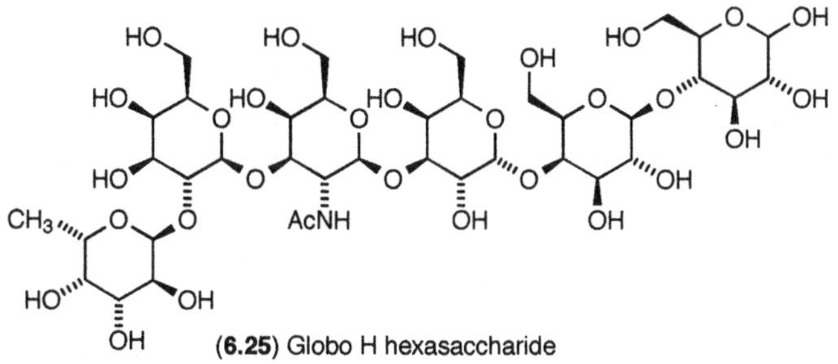

(6.25) Globo H hexasaccharide

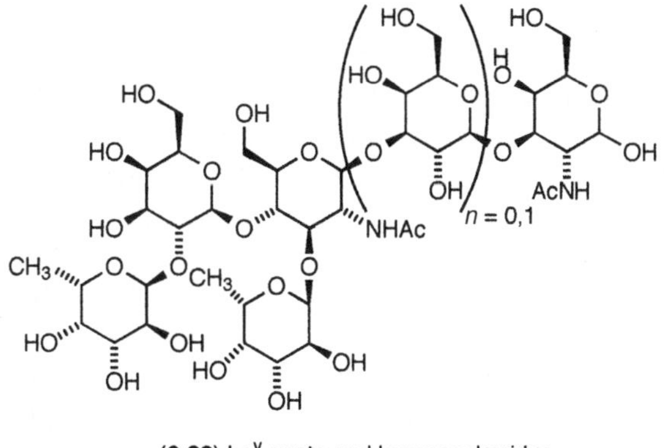

(6.26) Le[y] penta and hexasaccharides

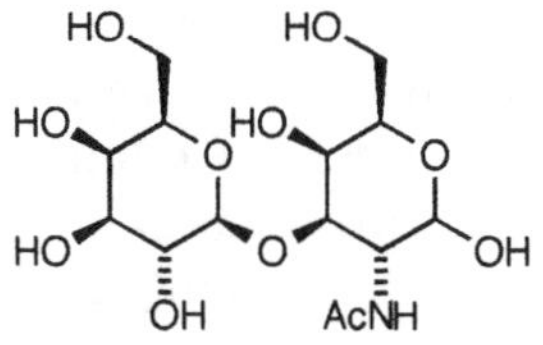

(6.27) TF disaccharide

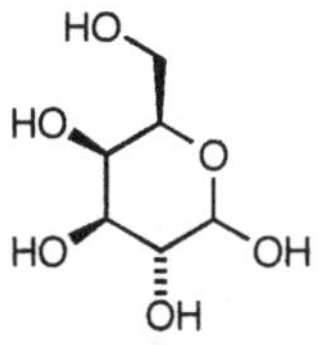

(6.28) Tn monosaccharide

proteins by reductive amination using sodium cyanoborohydride (Scheme 6.8(a)), or via a bifunctional linker group such as 4-(N-maleimidomethyl)-cyclohexane-1-carboxyl-hydrazide (**6.29**) (Scheme 6.8(b)).

A globo H-KLH conjugate and a Le^y-KLH conjugate, administered in concert with the immunological adjuvant QS21, have been used in clinical trials in breast and ovarian cancer patients, respectively. In general, the studies demonstrated that vaccination could generate an immune response with minimal toxicity. However, most of the antibody responses were modest. Second-generation vaccines have centred on constructing clustered carbohydrate epitopes as this may present the antigens more effectively to the immune system and augment the titres and duration of antibody response. While the antigen-to-KLH ratio of a glycoconjugate can readily be determined, it is not possible to establish the relative position of the carbohydrate antigens on the surface of the carrier protein. In order to ensure that the antigens form a cluster, they are assembled in the form of a glycopeptide and then this construct is conjugated to the immunogenic carrier protein to create a functional vaccine (Figure 6.3).

Attaching carbohydrates to a peptide is rarely successful mainly because it is difficult to effect glycosylation of amino acid residues with high anomeric stereoselectivity. The use of glycosylated amino acids as building blocks in the step-wise assembly is therefore a more reliable and efficient method of synthesizing glycopeptides and even bulky glycosyl amino acids can couple as successfully as ordinary non-glycosylated amino acids. Any method for activation and coupling used in the synthesis of peptides is, in principle, suitable for glycosylated amino acids but the carbodiimide/ hydroxybenzotriazole and the pentafluorophenyl ester methods predominate. Glycosidic bonds are labile towards strong acid and the Fmoc/t-butyl approach is more suitable for glycopeptide synthesis than the Boc/benzyl strategy which involves the use of HF (see Section 2.3.1 above). Carbohydrate hydroxyl functionality in glycosylated amino acids is often protected with acetyl groups which are easily removed (NH_3/MeOH or NaOMe/MeOH) at the end of the synthesis. Moreover, the acyl moiety can stabilize O-glycosidic bonds during acid-catalysed (TFA) treatment to remove side-chain protecting groups and, in a solid-phase synthesis, cleavage from the insoluble support.

The N-acetylgalactosamine-serine derivative (**6.30**) is the key building block used in the synthesis of a Tn antigen cluster and was prepared from the bromide **6.31** and N-Fmoc-serine benzyl ester **6.32** (Scheme 6.9). The fully protected (Tn-Ser)$_3$ glycopeptide **6.35** was assembled (Scheme 6.10) by using iterative standard (solution-phase) peptide couplings and then the benzyl ester at the C-terminus was removed by hydrogenolysis. Attachment of a linker to the free C-terminus, followed by removal of the acetyl protecting groups, afforded the fully deprotected glycopeptide **6.37**. Coupling of **6.37** to KLH using MBS (m-maleimimidobenzoyl-N-hydroxysuccinimide ester) (**6.38**), a heterobifunctional reagent which cross-links thiol groups with amino groups, gave the Tn cluster immunoconjugate component **6.39** of a potential cancer vaccine.

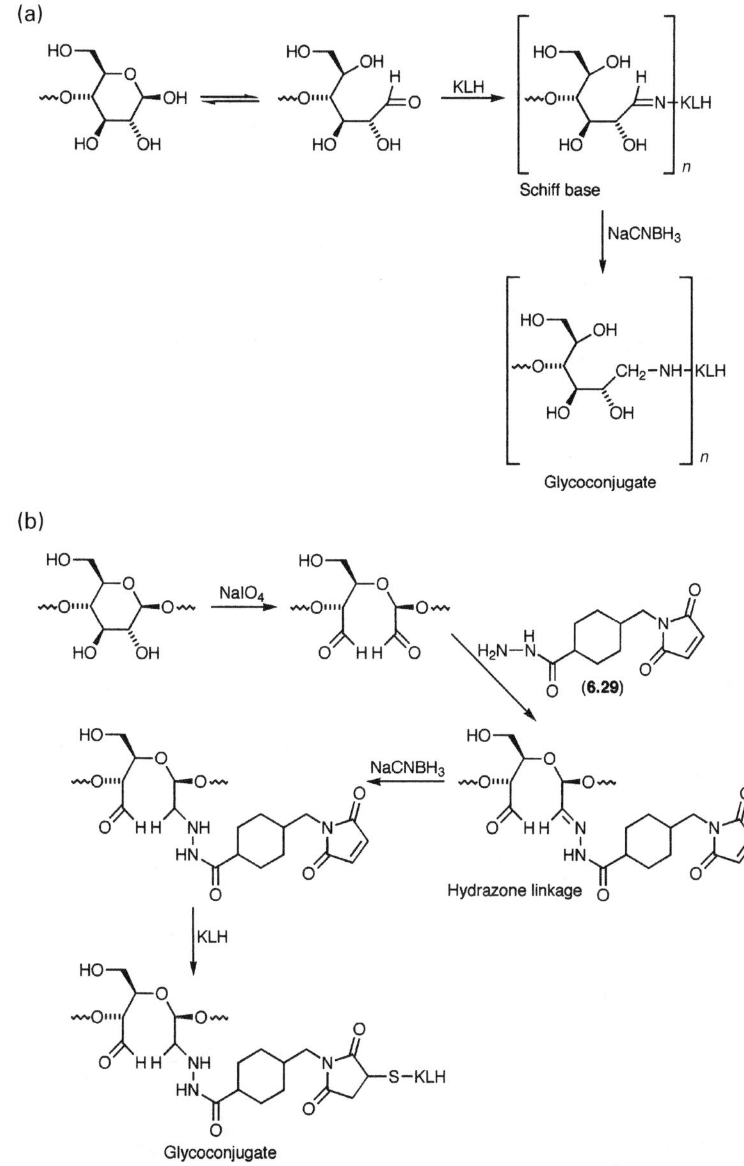

Scheme 6.8 Coupling of carbohydrates to proteins. (a) Direct conjugation. The open-chain aldehyde form of a reducing sugar (or created aldehyde, see (b)) reacts with amine-containing lysine residues on KLH to form a Schiff base. The reaction is carried out in the presence of a reducing agent such as sodium cyanoborohydride to convert the highly labile Schiff base to the more stable alkylamine linkage. (b) Indirect conjugation using a cross-linking reagent. Carbohydrates containing hydroxyl groups on adjacent carbon atoms can be oxidized with sodium periodate to create aldehyde groups. Reaction of the aldehyde with a hydrazine-containing cross-linking reagent forms a hydrazone linkage which can be reduced with sodium cyanoborohydride to form more stable linkages. The thiol-reactive maleimide end of the linker then reacts with cysteine groups on KLH to form thioether linkages and the glycoconjugate

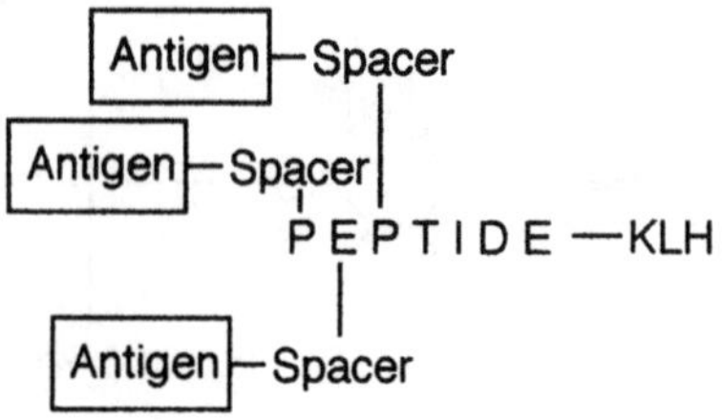

Figure 6.3 Antigen-clustered vaccines. The peptide serves as a scaffold for the antigenic carbohydrate epitopes. Protected glycoamino acids containing the oligosaccharide antigens are synthesized and then the glycopeptide is assembled from these building blocks. After deprotection the glycopeptide is conjugated to a suitable immunogenic carrier protein. Reprinted from *J. Am. Chem. Soc.*, 2001, **123**, 1890–1897, by permission of the American Chemical Society.

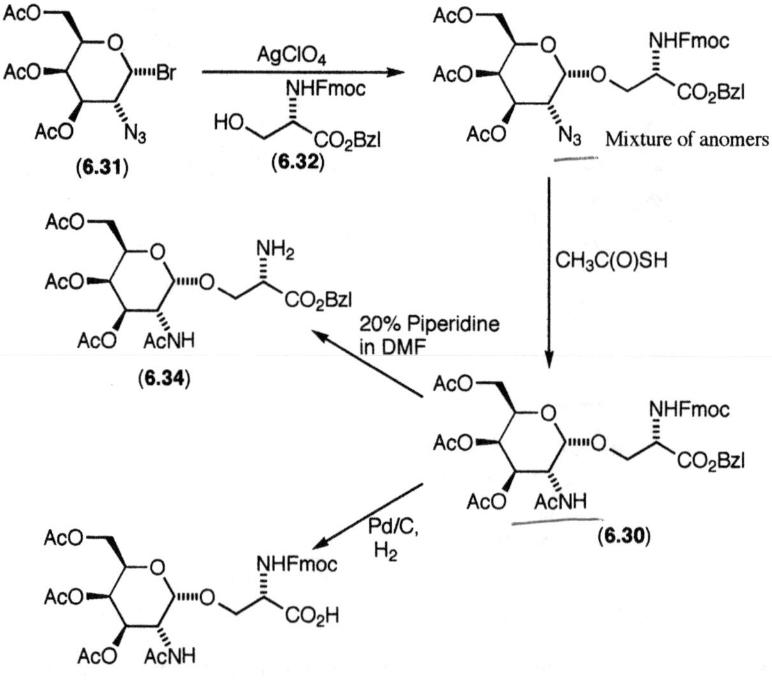

Scheme 6.9 Synthesis of protected GalNAc-serine residues for use as building blocks in glycopeptide synthesis. Glycosylation of bromide **6.31** with N-Fmoc-serine benzyl ester **6.32** gave a separable 4 : 1 α/β mixture of anomers. Reductive acetylation using thiolacetic acid produced the fully protected Tn antigen **6.30**. Hydrogenolysis gave the peracetylated acid **6.33**, while Fmoc removal gave **6.34**

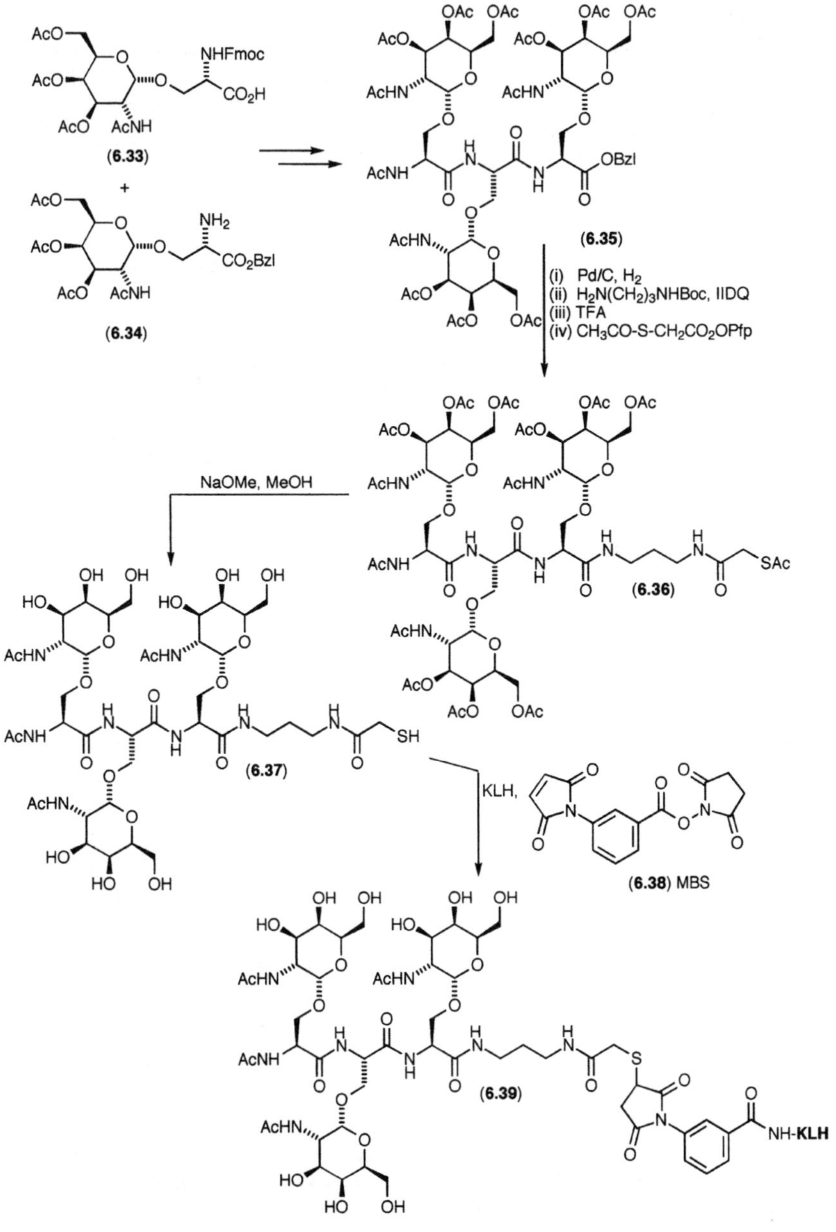

Scheme 6.10 Synthesis of a clustered Tn immunoconjugate. Protected Tn serines **6.33** and **6.34** were used as building blocks in the synthesis of the fully protected Tn tripeptide **6.35**. The glycosylated tripeptide thus obtained was subjected to hydrogenolysis to afford the acid which was coupled with *t*-butyl-*N*-(3-aminopropyl)carbamate in the presence of the activation agent 2-isobutoxy-1-isobutoxycarbonyl-1,2-dihydroquinoline (IIDQ). Removal of the Boc cap with TFA and coupling of the resulting amine with *S*-acetylthioglycolic acid pentafluorophenyl ester resulted in the fully protected glycopeptide **6.36**. Methanolysis cleaved the acetate esters to give the fully deprotected glycopeptide **6.37**, which was covalently linked with carrier protein KLH by using MBS

For the synthesis of his larger clustered glycopeptide vaccines, Danishefsky has developed a cassette methodology whereby orthogonally protected *N*-acetylgalactosamine is stereospecifically linked to a serine (or threonine) residue. With the required α-O-linkage in place, the glycoamino acid **6.40**, i.e. the cassette, can be used as a glycosyl acceptor and coupled to a saccharide

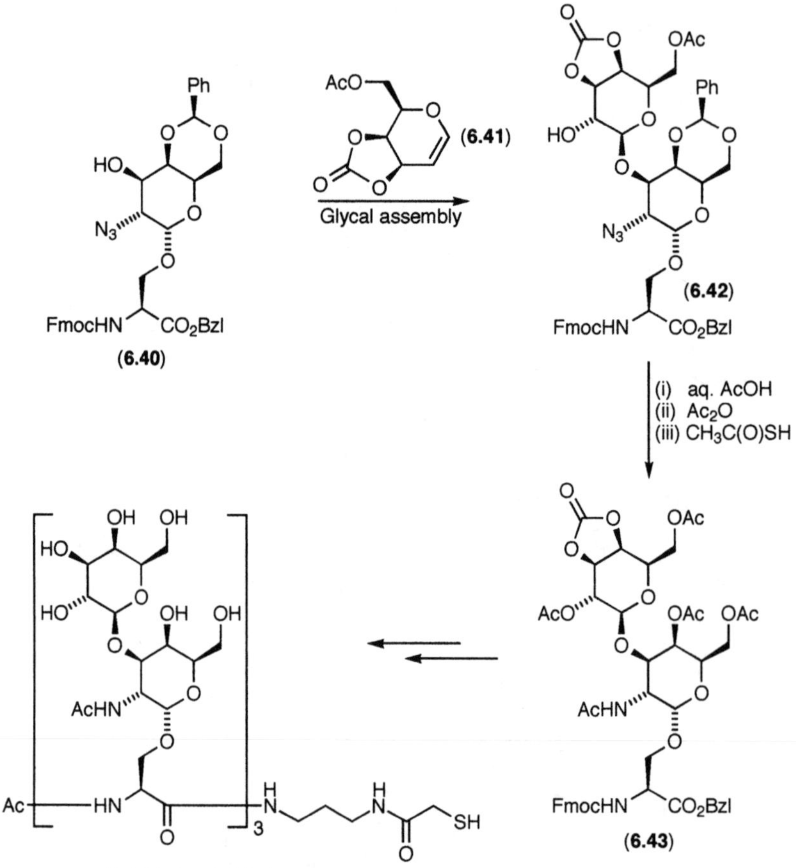

(6.44) Trimeric TF cluster

Scheme 6.11 Cassette-based synthesis of oligosaccharide-antigen-clustered glycopeptides. In the synthesis of the trimeric TF cluster **6.44**, glycal **6.41** which contains the TF specificity was reacted with the serine cassette **6.40** to afford the disaccharide **6.42** with the desired β-linkage. Removal of the benzylidene protecting group from positions 4 and 6, acetylation of free hydroxy groups and then reductive acetylation using thiolacetic acid produced the fully protected TF antigen–serine construct **6.43**. The latter was then advanced through the peptide assembly phase and a linker was attached to the C-terminus to afford the TF trimeric cluster **6.44**. In the synthesis of the Le^y cluster (not shown), the serine cassette was coupled in an analogous manner to the Le^y pentasaccharide glycal to produce the required serine-α-O-link to the complex carbohydrate domain of the Le^y antigen. Iterative peptide couplings of the Le^y serine construct and attachment of a linker to the resulting glycopeptide produced the clustered Le^y antigen suitable for conjugation to the carrier protein

donor in the synthesis of a target oligosaccharide–amino acid construct (Scheme 6.11). This is a better route to the heavily glycosylated amino acids than direct coupling of the complex carbohydrate antigens to the serine side-chain hydroxyl group, a reaction which is plagued by poor selectivity. By using the cassette strategy and glycal oligosaccharide synthesis techniques, disaccharide TF-serine and hexasaccharide Le^y-serine building blocks have been constructed and used in the preparation of clustered TF and Le^y glycopeptides and immunoconjugates. Vaccines based on clustered triads of antigens **6.26–6.28** e.g. (Le^y–Ser)$_3$, are under preclinical and clinical evaluation.

The first generation of antitumour vaccines and the more recent clustered designs are in essence constructions involving a single type of carbohydrate antigen displayed in polyvalent form. Vaccines containing several different carbohydrate antigens in a clustered format may have the potential to trigger a multifaceted immune response and result in a broader degree of immune protection against multiple cancers. Glycopeptides which each contain three different antitumour antigens, e.g. TF-Le^y-Tn and Tn-Le^y-Globo-H, have been synthesized and vaccines derived from these compounds are, at the time of writing (May 2001), under preclinical evaluation.

Further Reading

Textbook and review articles

- C. A. A. van Boeckel and M. Petitou, The Unique Antithrombin III Binding Domain of Heparin: A Lead to New Synthetic Antithrombotics, *Angew. Chem. Int. Ed. Engl.*, 1993, **32**, 1671–1690.
- S. J. Danishefsky and J. R. Allen, From the Laboratory to the Clinic: A Retrospective on Fully Synthetic Carbohydrate-Based Anticancer Vaccines, *Angew. Chem. Int. Ed. Engl.*, 2000, **39**, 837–863.
- S. J. Danishefsky and M. T. Bilodeau, Glycals in Organic Synthesis: The Evolution of Comprehensive Strategies for the Assembly of Oligosaccharides and Glycoconjugates of Biological Consequence, *Angew. Chem. Int. Ed. Engl.*, 1996, **35**, 1380–1419.
- S. David, *The Molecular and Supramolecular Chemistry of Carbohydrates: A Chemical Introduction to the Glycosciences*, Oxford University Press, Oxford, UK, 1997.
- B. G. Davis, Recent Developments in Oligosaccharide Synthesis, *J. Chem. Soc., Perkin Trans. 1*, 2000, 2137–2160.
- S. M. Hecht (Ed.), *Bioorganic Chemistry: Carbohydrates*, Oxford University Press, New York, 1999.

- J. Kihlberg, Glycopeptide Synthesis, in *Fmoc Solid Phase Peptide Synthesis: A Practical Approach*, W. C. Chan and P. D. White (Eds), Oxford University Press, Oxford, UK, 2000.
- J. H. Musser, P. Fügedi and M. B. Anderson, Carbohydrate-based Therapeutics, in *Burger's Medicinal Chemistry and Drug Discovery*, 5th Edn, Vol. 1, M. E. Wolff (Ed.), Wiley, New York, 1995, pp. 901–947.
- K. C. Nicolaou and H. J. Mitchell, Adventures in Carbohydrate Chemistry: New Synthetic Technologies, Chemical Synthesis, Molecular Design and Chemical Biology, *Angew. Chem. Int. Ed. Engl.*, 2001, **40**, 1577–1624.
- P. H. Seeberger and S. J. Danishefsky, Solid-Phase Synthesis of Oligosaccharides and Glycoconjugates by the Glycal Assembly Method: A Five Year Retrospective, *Acc. Chem. Res.*, 1998, **31**, 685–695.
- N. Sharon and H. Lis, Glycoproteins: Research Booming on Long-ignored, Ubiquitous Compounds, *Chem. Eng. News*, 1981 (March 30), 21–44.

Research publications

- J. R. Allen, C. R. Harris and S. J. Danishefsky, Pursuit of Optimal Carbohydrate-Based Anticancer Vaccines: Preparation of a Multiantigenic Unimolecular Glycopeptide Containing the Tn, MBr1, and Lewis^y Antigens, *J. Am. Chem. Soc.*, 2001, **123**, 1890-1897.
- M. Petitou, P. Duchaussoy, I. Lederman, J. Choay, P. Sinay, J.-C. Jacquinet and G. Toni, Synthesis of Heparin Fragments. A Chemical Synthesis of the Pentasaccharide O-(2-Deoxy-2-Sulfamido-6-O-Sulfo-α-D-Glucopyranosyl)-$(1\rightarrow4)$-O-(β-D-Glucopyranosyluronic   acid)-$(1\rightarrow4)$-O-(2-Deoxy-2-Sulfamido-3,6-Di-O-Sulfo-α-D-Glucopyranosyl)-$(1\rightarrow4)$-O-Sulfo-α-L-Idopyranosyluronic   acid)-$(1\rightarrow4)$-2-Deoxy-2-Sulfamido-6-O-Sulfo-D-Glucopyranose Decasodium Salt, A Heparin Fragment Having High Affinity for Antithrombin III, *Carbohydr. Res.*, 1986, **147**, 221–236.
- J. T. Randolph, K. F. McClure and S. J. Danishefsky, Major Simplifications in Oligosaccharide Syntheses Arising from a Solid-Phase Based Method: An Application to the Synthesis of the Lewis b Antigen, *J. Am. Chem. Soc.*, 1995, **117**, 5712–5719.
- P. Sinay, J.-C. Jacquinet, M. Petitou, P. Duchaussoy, I. Lederman, J. Choay and G. Toni, Total Synthesis of a Heparin Pentasaccharide Fragment Having High Affinity for Antithrombin III, *Carbohydr. Res.*, 1984, **132**, C5–C9.

Appendix

A.1 Amino Acids and Peptides

A.1.1 Abbreviations and structural representations

Most of the α-amino acids which are present in proteins can be represented by the general structure **A.1**. Chiral amino acids with this spatial arrangement of atoms have the L-configuration. In peptides and proteins, individual amino acids – the L-configuration is normally implied unless otherwise stated – are linked together by amide bonds to form chains (Figure A.1). A chain consisting of no more than 50 linked amino acids is usually referred to as a peptide, whereas if the number of amino acids exceeds 50 then the compound is usually referred to as a protein.

$$R \quad H$$
$$H_2N \quad CO_2H$$

(A.1)

There are 20 DNA encoded or proteinogenic amino acids (Figure A.2). Each amino acid has a three-letter and a one-letter abbreviation. The abbreviations can also be used to represent peptides and proteins. The three-letter abbreviations are usually used for the sequences of peptides, e.g. Lys–Ala–Gly = lysylalanylglycine (**A.2**), whereas the longer sequences of proteins are more commonly represented by the one-letter symbols. Using the three-letter symbols, the N and C-termini can be emphasized, and thus **A.2** can also be written as H–Lys–Ala–Gly–OH. Usually nothing is attached to the ends of the one-letter codes. Modified N- and C-termini can be indicated in the three-letter system, for example the derivative of **A.2** with acetyl and amide groups at the N and C-termini, respectively, would be given as Ac–Lys–Ala–Gly–NH$_2$. If the carboxylic acid group of the glycyl residue in **A.2** was transformed into a methyl ester, this would be abbreviated to –Gly–OMe while if it were reduced to an alcohol this would be written as –Gly–ol. Side-chains are not indicated in the abbreviations unless the functional group is substituted, e.g. the O-t-butylserine residue is shown as Ser(Bu^t). The SH groups in the cysteine side

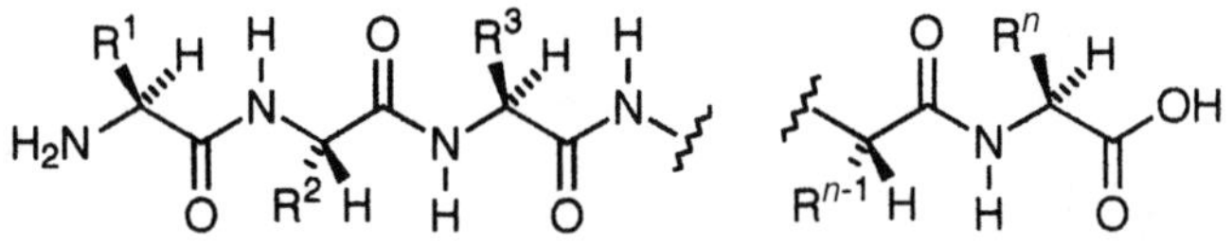

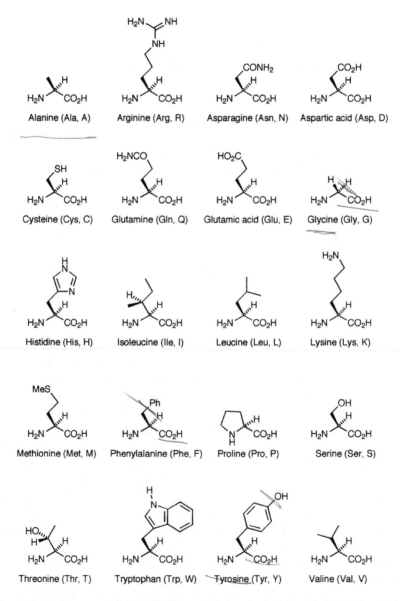

Figure A.1   Main chain or backbone structure of a polypeptide chain

Figure A.2   Structure of the 20 DNA-encoded amino acids and their abbreviations

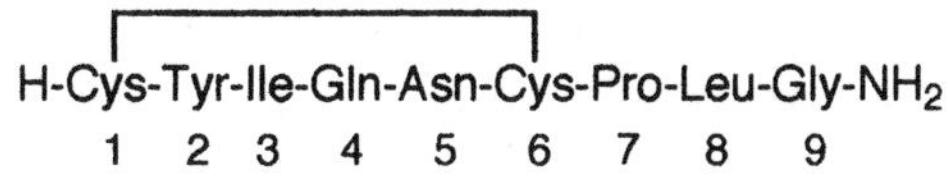
H-Cys-Tyr-Ile-Gln-Asn-Cys-Pro-Leu-Gly-NH2
1 2 3 4 5 6 7 8 9

(**A.3**) Oxytocin

chains can form a disulphide bond on oxidation and this is represented as in oxytocin (**A.3**). The residues of a peptide are by convention numbered from the N-terminal or left-hand end as shown by **A.3**.

A.1.2 Secondary structures

Rotation about amide bonds is inhibited by resonance; the lone pair of electrons of the nitrogen is delocalized on to the carbonyl oxygen and so the C–N bond has partial double-bond character (Figure A.3(a)). The amide bond is thus a rigid planar unit which restricts the polypeptide chain flexibility. The amide bond is usually found in a *trans*-configuration in peptides and proteins, although the less stable *cis*-isomer sometimes occurs (Figure A.3(b)). The conformation of a polypeptide chain is essentially described by the dihedral angles ϕ and ψ. Steric hindrance within the main chain and between main- and side-chain atoms further restricts the free rotation around the N–C_α and the C_α–C' bonds with the consequence that only a small range of ϕ, ψ combinations are 'allowed'. Given these geometric and steric constraints, it turns out that there are three low-energy arrange-

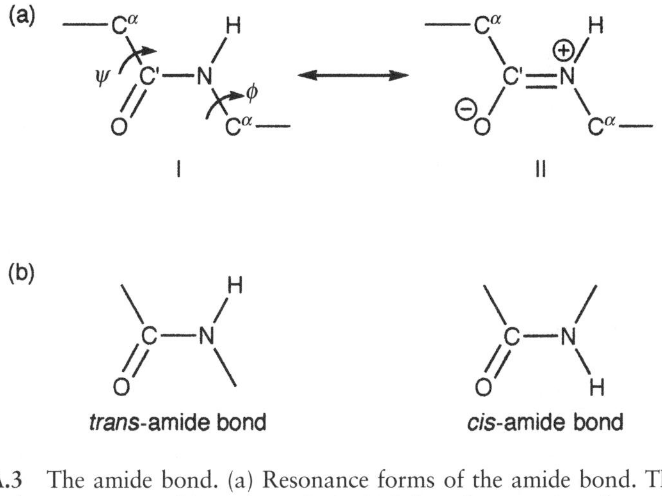
(a)
I
II
(b)
trans-amide bond
cis-amide bond

Figure A.3 The amide bond. (a) Resonance forms of the amide bond. The hybrid bond consists of structures I and II in the ratio 3 : 2; ϕ describes rotation about the N–C^α bond, while ψ describes rotation about the C^α–C' bond. (b) *Trans*- and *cis*-configurations of the amide bond

ments of the polypeptide backbone. These conformations, or secondary structures, are stabilized by specific patterns of hydrogen bonding.

(1) *α-Helix*   The carbonyl group of each amino acid residue forms a hydrogen bond with the NH of the fourth residue along in the sequence (Figure A.4). The C=O groups are parallel to the axis of the helix and point almost straight at the NH groups to which they are hydrogen-bonded. Approximately one-third of amino acid residues in proteins are in α-helices.

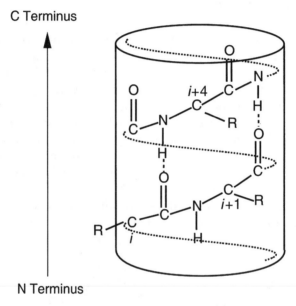

Figure A.4   Backbone structure of the α-helix. Repeated hydrogen-bonding between peptide carbonyl and NH groups of residues i and $i+4$ respectively leads to right-handed helices. Each hydrogen-bonded segment contains 13 atoms. The side-chains of the amino acids point away from the helix and there are 3.6 residues in each turn of helix

(2) *β-Sheet*   Side-to-side hydrogen bonding between fully-extended polypeptide chains. The β-sheet can take one of two forms, namely the parallel β-sheet in which all the chains are aligned in the same direction and the anti-parallel β-sheet in which the chains alternate in direction (Figure A.5). The repeating unit of a planar peptide bond linked to a tetrahedral carbon causes the sheets to be puckered or pleated with the side-chains extending away from the sheet to reduce steric crowding.

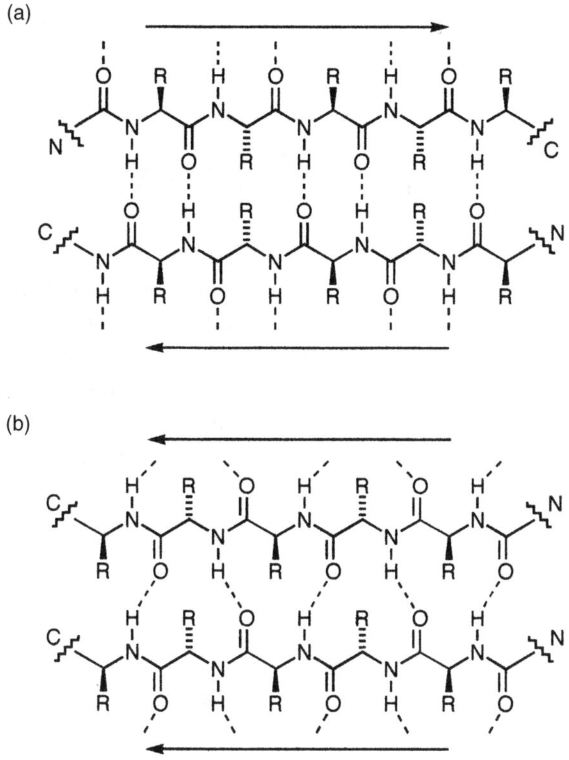

(a)

(b)

Figure A.5 Backbone structures of the (a) anti-parallel and (b) parallel β-pleated sheets

(3) *β-Turn* The anti-parallel β-structure can be achieved in a single polypeptide chain folding back on itself. Four amino acids are involved with
a hydrogen-bond formed between the carbonyl of the first residue and
the NH of the fourth, giving the equivalent of a 10-membered ring
(Figure A.6).

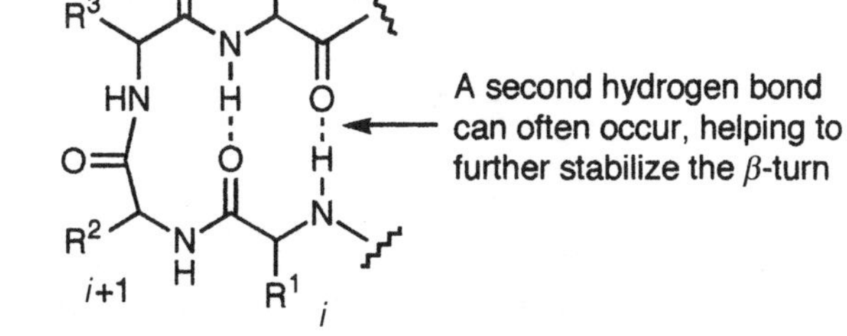

A second hydrogen bond
can often occur, helping to
further stabilize the β-turn

Figure A.6 Backbone structure of the β-turn

Protein three-dimensional structures are depicted in two dimensions by representing tracts of α-helices as cylinders and strands of β-sheet as large thick arrows.

A.2   Oligonucleotides, DNA and RNA

Nucleic acids consist of chains of nucleotides (Figure A.7). Each nucleotide consists of a nitrogen heterocyclic base, a pentose sugar and a phosphate group. There are two types of pentose sugars found in nucleic acids, namely 2′-deoxyribose (**A.4**) in DNA (deoxyribonucleic acid) and ribose (**A.5**) in RNA (ribonucleic acid). The bases are monocyclic pyrimidines or bicyclic purines. The pyrimidines are cytosine (C), uracil (U) and thymine (T), and the purines are adenine (A) and guanine (G) (Figure A.8). DNA contains A, G, T and C,

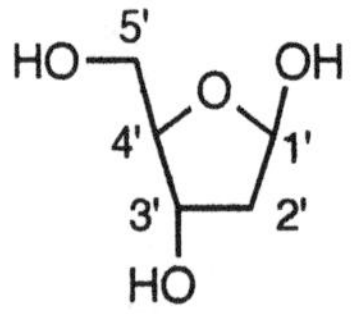

(A.4) 2′-Deoxyribose 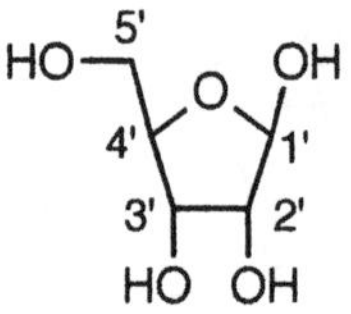   **(A.5)** Ribose

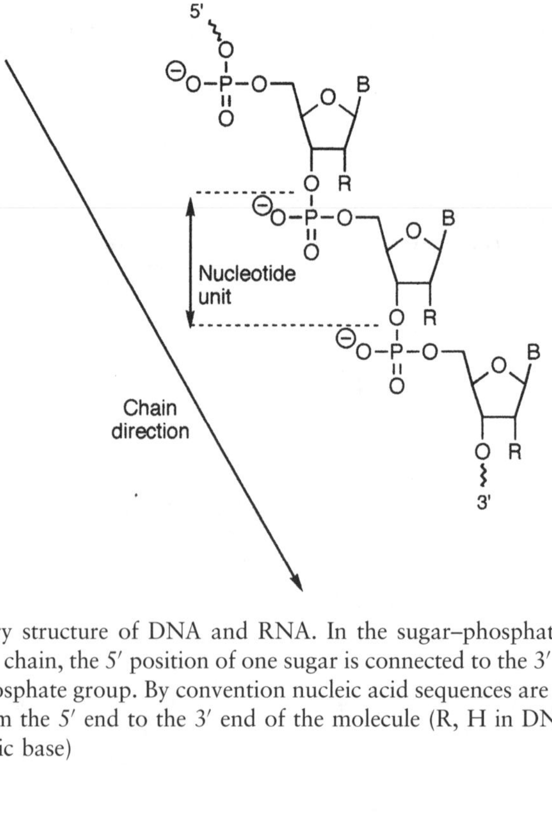

Figure A.7   Primary structure of DNA and RNA. In the sugar–phosphate backbone of each polynucleotide chain, the 5′ position of one sugar is connected to the 3′ position of the next sugar via a phosphate group. By convention nucleic acid sequences are always written in the direction from the 5′ end to the 3′ end of the molecule (R, H in DNA, and OH in RNA; B, heterocyclic base)

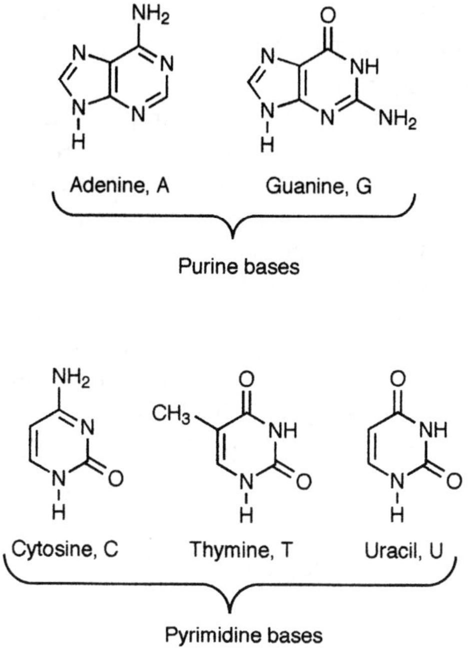

Figure A.8   Structures of the five purine and pyrimidine bases of nucleic acids

while RNA also has A, G and C but with T being replaced by U. The base-sugar moiety is called a nucleoside (Table A.1), while the base-sugar-phosphate moiety is called a nucleotide.

RNA consists of a single polynucleotide chain, whereas DNA is composed of two polynucleotide chains arranged in a double helix (Figure A.9). The two strands of DNA run in opposite directions, i.e. are anti-parallel. Projecting inwards from the two sugar–phosphate backbones are the heterocyclic bases which exhibit specific pairing: guanine in one chain always pairs with cytosine in the other chain and adenine always pairs with thymine. The two strands are held together by hydrogen bonds (Figure A.10). The $5' \rightarrow 3'$ strand is the coding or sense strand and the sequence is in the same $5' \rightarrow 3'$ orientation as

Table A.1   Heterocyclic bases and their equivalent nucleosides

Base	Nucleoside	
	DNA	RNA
Adenine	Deoxyadenosine	Adenosine
Guanine	Deoxyguanosine	Guanosine
Cytosine	Deoxycytidine	Cytidine
Thymine	Deoxythymidine	—
Uracil	—	Uridine

the resultant RNA and so can be 'read' exactly as the RNA sequence can (except for the substitution of T for U, see below). The 3′ → 5′ strand is called the template or anti-sense strand (Figure A.11).

The order of the bases attached to the sugar–phosphate backbone of DNA dictates the genetic code. Genetic information contained in DNA is transmitted to messenger RNA (mRNA) by a process referred to as transcription which involves the DNA being used to order complementary sequences of bases in the mRNA. Genetic information in mRNA then directs protein synthesis in a process referred to as translation (Scheme A.1). Genetic information within the DNA molecule is stored in the form of triplet codes, i.e. a sequence of three nucleotide bases determines the formation of one amino acid. The triplet of bases which codes for one amino acid is called a codon. Since the genetic code is actually read from the mRNA it is often represented in terms of the four bases on RNA, namely, A, G, C and U (Table A.2).

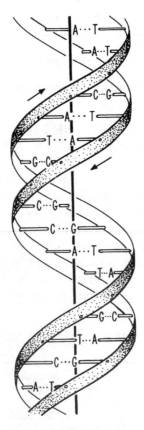

Figure A.9 Schematic drawing of the DNA double helix. The sugar-phosphate backbones run at the periphery of the helix in opposite directions. Base pairs (A–T and G–C), drawn symbolically as bars between chains, are stacked along the centre of the helix. (Used with permission).

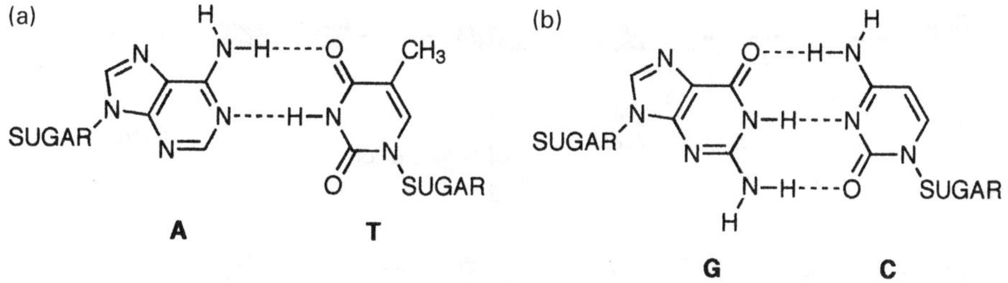

Figure A.10 DNA bases showing hydrogen-bonding in Crick–Watson base pairs: (a) A–T; (b) G–C

A.3 Carbohydrates

The most common carbohydrate monomers **A.6** contain a tetrahydropyran ring, three secondary OH groups, a side-chain carrying a primary OH function and a hemiacetal OH at C-1. These monosaccharides differ in the relative configurations of the four chiral centres at C-2, C-3, C-4 and C-5. For example, of the 2^4 different stereoisomers of the molecule with the general structure **A.6**, D-glucose (**A.7**) and D-mannose (**A.8**) have the opposite configurations at C-2, while D-galactose (**A.9**) is the epimer of D-glucose at C-4. By convention, the D-series of isomers has the D-glyceraldehyde configuration (*R*) at C-5, while the L-series has the L- or *S*-glyceraldehyde configuration. The hemiacetal carbon at C-1 is distinguished from the others by its unstable configuration and is referred to as the anomeric carbon. The α- and β-anomers interconvert via an open-chain form carrying, at C-1, the latent aldehyde group and, at C-5, a hydroxyl group (Scheme A.2).

While the D-sugars predominate in nature, some key units of the glycoprotein oligosaccharide chains belong to the L-series, with the partially deoxygenated sugar L-Fucose (**A.10**) being the most notable example. Oligosaccharides can also contain simple derivatives of the monosaccharides, e.g. replacing the hydroxyl at C-2 by an acetamido group gives *N*-acetylhex-

```
              Met Gly His Phe Tyr
     { 5'------ATG GGG CAC TTC ACA ------3' ←— Coding (sense) strand
DNA  {
     { 3'------TAC CCC GTG AAG TGT ------5' ←— Template (antisense) strand

RNA    5'------AUG GGG CAC UUC ACA ------3'

Direction of transcription —→
```

Figure A.11 Definition of DNA strands. The upper strand corresponds in sequence to the RNA that is made (except for the substitution of U in RNA for T in DNA). The lower strand is the antisense strand and acts as the template for transcription. By convention, when a DNA sequence is specified it is presented as the coding strand; thus the sequence ATG GGG CAC TTC ACA encodes the pentapeptide Met-Gly-His-Phe-Tyr

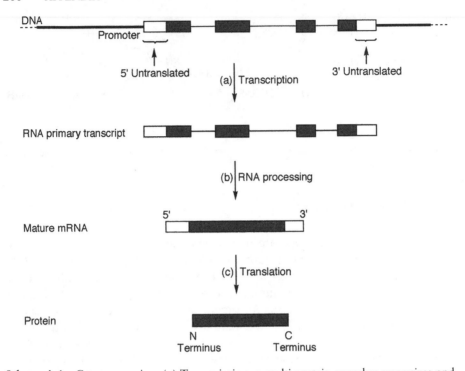

Scheme A.1 Gene expression. (a) Transcription: a multi-protein complex recognizes and binds to the promoter site; a nucleotide sequence near one end of the gene. The complex moves along the chain and causes the DNA strands to separate within a short region (10 to 20 base-pairs). RNA polymerase, one of the components of the protein complex, moves into the gene and synthesizes a new chain by linking together RNA nucleotides. The nucleotides are added to the 3′-OH end of the RNA chain which grows in the 5′ to 3′ direction. The DNA strand serving as the template is therefore traversed from its 3′ to its 5′ end. This means that the process of transcription produces the complement of the template strand, i.e. the RNA strand has the same sequence as the strand of DNA that is not transcribed (except with Us instead of Ts). (b) RNA processing: the initial product of transcription is an RNA precursor which is longer than the mature, translatable product. The additional nucleotides in these longer RNAs are introns or intervening sequences and they are removed by splicing, a process in which a complex tertiary structure is formed and the intron sequence is eliminated, thus bringing the coding sequences (exons) together. Mature mRNA has all the introns removed and usually has a 5′-untranslated region, a coding region and a 3′-untranslated region. The untranslated regions are, as the name suggests, not translated but contain sequences which affect the translational efficiency and stability of the mRNA. (c) Translation: mRNA migrates out of nucleus into the cytoplasm where it becomes associated with the ribosomes which are the site of protein synthesis. The amino acids which will be linked together to form the protein are attached to a family of transfer RNA (tRNA) molecules, each of which recognizes, by complementary base-pairing interactions, particular three-nucleotide sequences (codons) on the mRNA. The ribosome moves along the mRNA molecule, translating the nucleotide sequence into an amino acid sequence one codon at a time using tRNA molecules to add amino acids to the growing end of the polypeptide chain

Table A.2 The triplet codons of the genetic code. For example, the codons CAC and UGG are translated into histidine and tryptophan, respectively

1st Base	2nd Base				3rd Base
(5' end)	A	G	C	U	(3' end)
A	Lys	Arg	Thr	Ile	A
	Lys	Arg	Thr	Met	G
	Asn	Ser	Thr	Ile	C
	Asn	Ser	Thr	Ile	U
G	Glu	Gly	Ala	Val	A
	Glu	Gly	Ala	Val	G
	Asp	Gly	Ala	Val	C
	Asp	Gly	Ala	Val	U
C	Gln	Arg	Pro	Leu	A
	Gln	Arg	Pro	Leu	G
	His	Arg	Pro	Leu	C
	His	Arg	Pro	Leu	U
U	STOP	STOP	Ser	Leu	A
	STOP	Trp	Ser	Leu	G
	Tyr	Cys	Ser	Phe	C
	Tyr	Cys	Ser	Phe	U

osamines (e.g. **A.11** and **A.12**), while oxidation at C-6 leads to the uronic acids D-glucuronic acid (**A.13**) and L-iduronic acid (**A.14**).

There are also five-carbon sugars of which D-ribose and 2-deoxy-D-ribose, key constituents of ribonucleic acids and deoxyribonucleic acids respectively, are the most important (see Section A.2 above).

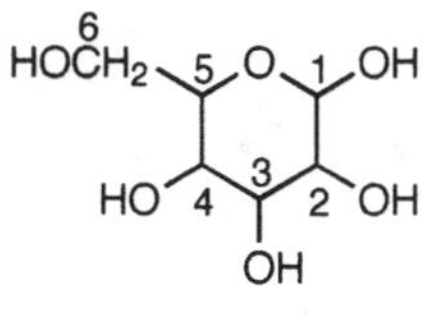

(**A.6**)

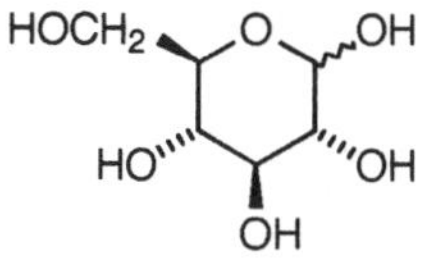

(**A.7**) D-Glucose (Glc)

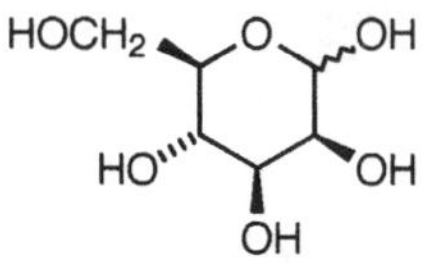

(**A.8**) D-Mannose (Man)

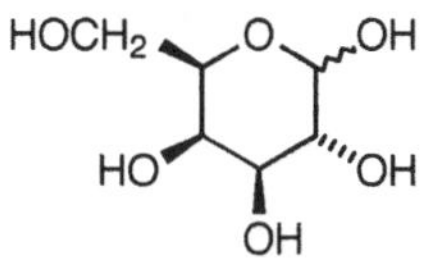

(**A.9**) D-Galactose (Gal)

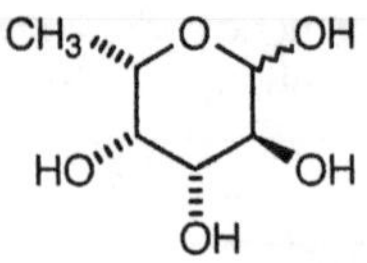

$$CH_3\text{'''}\underset{HO\text{'''}}{\overset{O}{\bigcirc}}\underset{OH}{\overset{OH}{\text{'''OH}}}$$

(A.10) L-Fucose (Fuc)

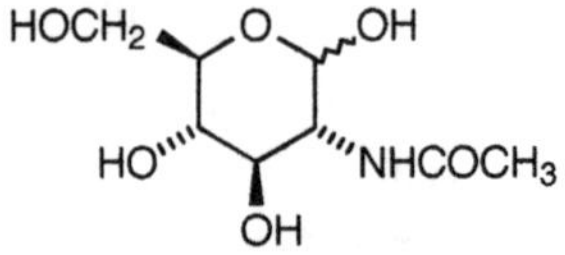

$$HOCH_2\underset{HO\text{'''}}{\overset{O}{\bigcirc}}\underset{OH}{\overset{OH}{\text{'''NHCOCH}_3}}$$

(A.11) N-Acetylglucosamine
(GlcNAc)

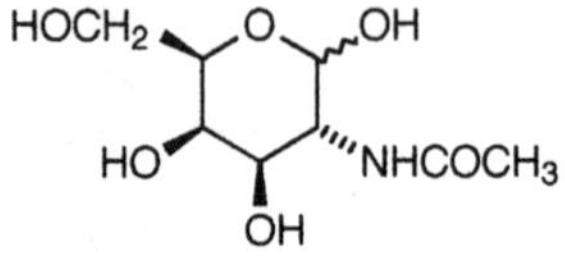

$$HOCH_2\underset{HO\text{'}}{\overset{O}{\bigcirc}}\underset{OH}{\overset{OH}{\text{'''NHCOCH}_3}}$$

(A.12) N-Acetylgalactosamine
(GalNAc)

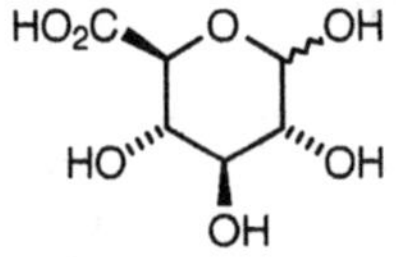

$$HO_2C\underset{HO\text{'''}}{\overset{O}{\bigcirc}}\underset{OH}{\overset{OH}{\text{'''OH}}}$$

(A.13) D-Glucuronic acid

$$HO_2C\text{'''}\underset{HO\text{'''}}{\overset{O}{\bigcirc}}\underset{OH}{\overset{OH}{\text{'''OH}}}$$

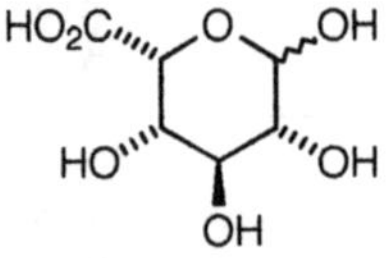

(A.14) L-Iduronic acid

Monosaccharides are joined together by an acetal linkage to form poly-saccharide chains through the reaction of the hemiacetal of one unit with a hydroxyl group of another unit. The acetal or glycosidic link is indicated by the symbol $(1 \to m)$ where $m = 1-4$ or 6. The anomeric descriptor (α or β) is placed after the abbreviation for the sugar. Thus, the H antigen, which is characteristic of individuals with blood group O (see Section 6.4.1 earlier), is written as L-Fuc-β-(1→2)-D-Gal.

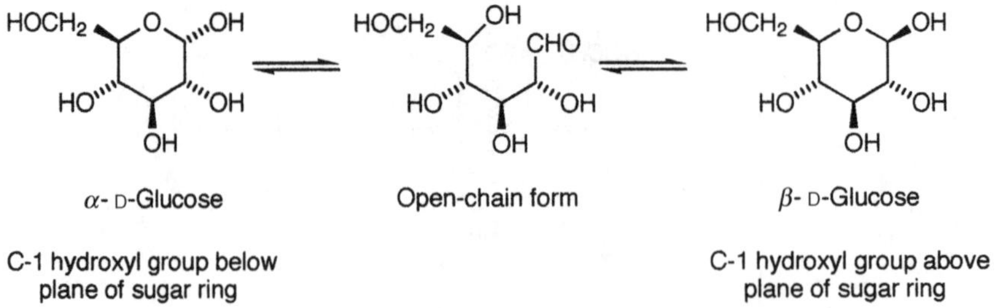

$$HOCH_2\underset{HO\text{'''}}{\overset{O}{\bigcirc}}\underset{OH}{\overset{\text{'''OH}}{\text{'''OH}}} \rightleftharpoons HOCH_2\underset{HO\text{'''}}{\overset{OH\,CHO}{\bigcirc}}\underset{OH}{\overset{}{\text{'''OH}}} \rightleftharpoons HOCH_2\underset{HO\text{'''}}{\overset{O}{\bigcirc}}\underset{OH}{\overset{OH}{\text{'''OH}}}$$

α- D-Glucose   Open-chain form β- D-Glucose

C-1 hydroxyl group below
plane of sugar ring

C-1 hydroxyl group above
plane of sugar ring

Scheme A.2   Interconversion of α- and β-anomers of D-glucose

Glossary

Adjuvant A substance that when mixed with an antigen enhances the immune response.

Adoptive T-cell immunotherapy Passive immunotherapy with cells, usually following *ex vivo* stimulation of the cells to enhance anti-tumour reactivity.

Aetiology Cause or causes of disease.

Allograft A tissue transplant (graft) between two genetically non-identical members of a species.

Anaphylactic shock A severe and sometimes fatal acute allergic reaction in which there is a sudden, generalized shock and collapse.

Anchimeric Facilitation, by one part of a molecule, of a reaction that occurs at a different part of the same molecule.

Annealing *See* Hybridize.

Anomeric effect The preference of an electronegative substituent at the anomeric centre of a carbohydrate for adopting the axial orientation, rather than the expected, more sterically favoured, equatorial position.

Antibody A serum protein which is formed in response to an antigenic stimulus and reacts specifically with this antigen; part of the body's defence against infection.

Antibody-dependent cellular cytotoxicity A reaction in which an antibody-coated target cell is directly killed by specialized killer cells (natural killer cells and macrophages) bearing receptors for the Fc portion of the coating antibody.

Antigen A substance that the body recognizes as foreign and which can stimulate an immune response.

Antigen-binding site The part of an antibody molecule or T-cell receptor that binds antigen specifically.

Antigenic determinant *See* Epitope.

Antigen-presenting cell (APC) A specialized type of cell, bearing cell-surface MHC class II molecules, involved in processing and presentation of antigen to T cells.

Antisense A strand of DNA that has the sequence complementary to mRNA.

Antiserum (pl. antisera) The serum of an animal that contains a heterogeneous collection of antibodies specific for the molecule used for immunization.

Apoptosis Form of cell death controlled by an internally encoded suicide programme.

Attenuated Reduction of the virulence of a bacterium or virus by physical or chemical treatment or through selection of variants.

AUG Codon for the first amino acid in protein sequences, which is methionine in eukaryotes and formylmethionine in prokaryotes (fMet is often removed post-translationally).

Autoantigen A normal tissue constituent that evokes an immune response.

Autoimmune disease Disorder in which the immune system identifies normal body components as antigens and produces antibodies to them.

Autologous Derived from the same individual.

B cells Cells which secrete antibodies.

Bioavailability Proportion of the administered dose which is absorbed into the bloodstream; may be low when drugs are given orally.

Burkitt's lymphoma Malignant tumour of the lymphatic tissue, most commonly affecting children and usually found in central Africa. Rapidly growing malignancy. The Epstein–Barr virus has been implicated as a causative agent.

Bystander effect Killing or damage of cells neighbouring those directly targeted by a specific therapeutic procedure.

Capsid Protein coat of a virus particle.

Carcinoma A malignant tumour causing death, i.e. cancer.

CD Cluster of differentiation. Cell surface molecules with which antibodies react and which are used to classify lymphocytes. Each of the CD proteins was originally defined as a T-cell 'differentiation antigen'.

cDNA A single-stranded DNA complementary to an mRNA template and synthesized from it by *in vitro* reverse transcription.

Chimeric antibody A recombinant antibody that has the Fab fragment of one species (usually mouse) fused with the Fc fragment from another species (usually human).

Clinical trial An experiment on humans carried out in order to evaluate the efficacy of one or more potentially beneficial therapies. In general, clinical trials are characterized by three phases, referred to as phase I, II or III trials:

- Phase I Clinical trial on normal volunteers, designed to determine the biological activities and range of toxicity or other safety factors of a new treatment (usually a drug).

- Phase II   Clinical trial on a small group of patients, designed to determine the effectiveness of the given regimen in treating the disorder in question.

- Phase III   Clinical trial using a large sample of patients, designed to compare the overall course of their disorder under the new treatment with its course untreated or treated with standard therapies previously used.

In cancer clinical trials, phase I studies are toxicity trials to establish the maximum tolerated dose. They involve patients who are unresponsive to therapies that are believed to be beneficial. Side-effects are more acceptable in cancer studies than in trials targeting less serious illnesses. Phase II cancer trials are designed to determine activity in specific tumour types. Phase III studies are randomized trials with one or more experimental therapies compared with the best standard therapy for that disease.

Clone   (1) Noun: one of a collection of cells or vectors containing identical genetic material. (2) Verb: the act of duplicating genetic material within a vector.

Coding strand (of DNA)   Strand in duplex DNA that contains the same sequence of bases as mRNA. Also known as the sense strand.

Codon   A sequence of three adjacent nucleotides (triplet) which codes for one amino acid or chain termination.

Complement   A series of serum proteins involved in the mediation of immune reactions; the complement cascade is triggered by the interaction of an antibody with a specific antigen.

Complementarity-determining region (CDR)   The portion of the antibody variable region that binds to antigen. It is composed of hypervariable sequences, with three from each of the heavy and light chains. The six CDR loops together form the site involved in antigen binding.

Conformational epitope   An epitope comprised of contiguous but physically discontinuous components of the immunogenic molecule.

Crohn's disease   Chronic inflammatory disease of the gastrointestinal tract.

Cytokines   Soluble substances secreted by cells, which have a variety of effects on other cells. *See also* Interleukins.

Cytoplasm   The jelly-like substance that surrounds the nucleus of a cell and carries structures within which most of the cell's life processes take place.

Cytotoxic T cells   A sub-set of T cells that kill virally infected or cancerous cells of the body.

Denaturation   Unfolding of a protein to generate an inactive form; also the conversion of double-stranded DNA into two separate strands.

DNA cloning   Production of many identical copies of a defined DNA fragment.

DNA ligase   Enzyme used to join DNA molecules.

DNA polymerase   An enzyme that catalyses the synthesis of double-stranded DNA from single-stranded DNA.

Downstream   DNA sequences after the start of a gene.

Efficacy   The ability of a drug to produce the desired therapeutic effect.

Emphysema   Accumulation of air in certain parts of the body, especially the lungs.

Endocrine system   Cells, tissues and glands which secrete hormones directly into the bloodstream to act at distant sites.

Endocytosis   Process by which extracellular material is taken up by a cell.

Endogenous   Arising within or derived from the body.

Endonuclease   *See* Restriction endonuclease.

Enhancer   A 50–150 base pair sequence of DNA that increases the rate of transcription of coding sequences.

Epitope   A part of the antigen molecule which binds to an antibody-combining site or to a receptor on T cells and which elicits an immune response.

Eukaroytes   Higher organisms with a well-defined nucleus.

Exon   A nucleotide sequence in a gene that codes for a section of mature mRNA and therefore encodes the protein sequence.

Expression   To 'express' a gene is to cause it to function. A gene which encodes a protein will, when expressed, be transcribed and translated to produce that protein.

Ex vivo   Outside the living body; removal of tissue for treatment after which it is returned to the original site.

Gaucher's disease   Disease characterized by abnormal storage of lipids, particularly in the spleen, bone marrow and liver; results in enlargement of the spleen and the liver and anaemia.

Gene   A DNA sequence involved in the production of an RNA or protein molecule as the final product.

Genome   All of the genetic material in the chromosomes of an organism.

Germ line   Genes in germ cells (sperm and ovum) as opposed to somatic cells.

Haematopoietic   Blood-forming.

Haematopoietic stem cells   The cells for which all classes of blood cells are derived.

Helper T cells   A class of T cells; some release cytokines that stimulate antibody production by activated B cells. They also help in the differentiation of other T cells such as cytotoxic T cells.

Hodgkin's disease   A malignant lymphoma characterized by painless, progressive enlargement of lymphatic tissue.

Hormone A chemical substance secreted by cells of the endocrine system and transported through the bloodstream to target cells with appropriate receptors where it elicits a response.

Host An organism that provides the life-support system for another organism, virus or plasmid.

Human anti-mouse antibodies (HAMAs) Antibodies formed by patients following repeated administration of monoclonal antibodies of murine origin.

Humanization The process by which a non-human antibody is made as near human as possible.

Humanized antibody A recombinant antibody whose framework is composed of human immunoglobulin but whose CDR regions are of murine origin.

Hybridize To establish base-pairing between complementary strands of nucleic acid.

Hybridoma A clone of cells derived from the fusion of an antibody-secreting B cell and a myeloma cell that can proliferate indefinitely in a test tube.

Hypothalamus A wedge-shaped mass of tissue which lies at the lowest part of the mid-brain just above the pituitary.

Immunodeficiency Decrease in the immune response which results from the absence or a defect of some component of the immune system.

Immunoglobulin *See* Antibody.

Inducer A compound which stimulates expression of some genes.

Inflammatory response A non-specific defence mechanism triggered by antigens or tissue damage; results in swelling, redness, warmth and pain.

Interleukin Secreted peptides and proteins that mediate intercellular signals between white blood cells but do not bind antigen. Those secreted by lymphocytes (T cells and B cells) are also called lymphokines. Most of the interleukins have many more sources, targets and actions and are therefore more accurately called cytokines.

Intron (= Intervening sequence). A non-coding section of a gene; it is transcribed into RNA but removed before mature mRNA proceeds to protein synthesis. Bacterial mRNA does not contain introns.

Invariant A region of a protein that does not alter in sequence between genetic variants.

In vitro **(lit. 'in glass')** Refers to any biological process occurring outside of the living cell.

In vivo Refers to any biological process occurring within the living cell.

Isoform One of the multiple forms of the same protein differing in their primary structure but which possess the same function.

Isostere   Substitute for the amide bond.

Leukaemia   A group of malignant diseases of the bone marrow and other blood-forming organs.

Liposomes   Small spherical artificial lipid membranes.

Lymphocytes   Variety of white blood cell formed in lymphatic tissue (e.g. lymph nodes, spleen and bone-marrow). They are involved in immunity and can be subdivided into B lymphocytes (B cells) which produce circulating antibodies and T lymphocytes (T cells) which are primarily responsible for cell-mediated immunity.

Lymphoma   Malignant proliferation of lymphoid tissue. *See also* Burkitt's lymphoma and non-Hodgkin's lymphoma.

Macrophage   Large scavenger cell.

Major histocompatibility complex   A set of genes encoding cell-surface glyco-proteins (MHC class I and class II) that are involved in the presentation of antigens to antigen-specific T cells. MHC is also known as human leukocyte antigen or HLA system. These glycoproteins also play a major role in transplantation rejection.

Malignant   Tumour that invades and destroys the tissue in which it originates and can spread to other sites of the body via the bloodstream and lymphatic tissue (*see* Cancer).

Melanoma   Tumour arising from the cells, most commonly in the skin, which produce melanin.

Metastasis   The development of secondary tumours distant from the original site of malignancy.

Monoclonal antibody   Pure antibody produced by a single clone of cells derived by fusion of a normal antibody-producing B cell with an immortalized myeloma cell.

mRNA   The RNA which contains sequences coding for a protein; used only for a mature transcript with all introns removed, rather than the primary transcript in the nucleus.

Mutation   An error in base sequence that is carried along in DNA replication.

c-*myb*   Protooncogene that encodes a DNA binding protein (c-Myb) which is involved in the differentiation of haematopoietic cells.

c-*myc*   Protooncogene encoding a transcription factor (c-Myc) that has a key role in cell proliferation and differentiation. Deregulation by overexpression of c-*myc* blocks exit from the cell cycle and leads to a population of continuously proliferating cells.

Myeloma   A tumour of plasma cells, generally secreting a single monoclonal antibody.

Myocardial infarction Destruction of an area of heart muscle due to blockage of the coronary artery.

Nasopharyngeal carcinoma Tumour affecting the nasal cavity.

Natural killer cells Antigen-non-specific cytotoxic lymphocytes that are able to kill virus infected cells and certain types of cancer cell.

Neoplasm (= Tumour). Any abnormal growth of tissue.

Neuroblastoma Malignant pediatric brain tumour.

Neurotransmitter A chemical messenger that travels between a neuron and a neighbouring neuron or other target cell to transmit a nerve impulse.

Non-Hodgkin's lymphoma Similar to Hodgkin's disease but is initially more widespread.

Oncogene Cancer gene. Genes which are capable, under certain conditions, of inducing a normal cell to be transformed into a malignant state. A prefix 'v-' indicates that the gene is derived from a virus.

Operator A region on DNA at which a repressor protein binds to prevent transcription of an adjacent gene.

Operon Complete unit of bacterial expression consisting of (a) regulator gene(s), control elements (promoter and operator) and adjacent structural gene(s). The *lac* operon is responsible for the entry and metabolism of lactose and consists of genes coding for three enzymes flanked by a repressor and a promoter region to control expression.

Origin of replication A nucleotide sequence present in a plasmid that serves as a start signal for DNA replication.

Orthogonal protecting groups Combinations of protecting groups such that each group can be removed by different chemical mechanisms; therefore, the groups can be removed in any order and in the presence of other groups.

Pancreas A gland, approximately 15 cm long, situated behind the stomach.

Parenteral Administration of drug by means other than through the mouth.

Passive immunization Immunization of an individual by the transfer of antibody synthesized in another individual.

Pathogen An agent which causes disease.

pBR322 A standard plasmic cloning vector.

Peptidase *See* Protease.

Peptide mimetic (= Peptidomimetic). Chemical structure derived from a bio-active peptide which imitates natural molecules.

Periplasm Space between the inner and outer membranes of a bacterial cell.

Phage   A virus that infects bacteria, used in the laboratory as a cloning vector.

Phagocytosis   The engulfment and digestion of a particle or a micro-organism by cells such as macrophages and neutrophils.

Pharmacodynamic   The interaction of drugs with cells, i.e. the biochemical and biophysiological effects of drugs and the mechanisms of their actions; what a drug does to the body.

Pharmacokinetic   Movement of a drug within the body encompassing its absorption, distribution, metabolism and elimination, i.e. what the body does to a drug.

Pharmacophore   The group of atoms in a molecule which is responsible for its biological activity.

Pituitary   A pea-sized body located at the base of the brain and attached by a stalk to the hypothalamus. It is divided into two regions, i.e. anterior and posterior.

Plasma   Fluid component of whole blood.

Plasmid   A small circular piece of DNA found inside bacterial cells which is capable of autonomous replication.

Polyclonal antibody   Antibodies derived from more than a single clone of cells.

Polymerase chain reaction   *In vitro* amplification of a specific piece of DNA by repeated rounds of oligonucleotide binding and extension.

Polymorphic gene   A gene that exists in multiple forms and results in variants of gene product.

Post-translational modification   Various modifications (e.g. phosphorylation, glycosylation and proteolytic cleavage) of proteins which occur after their synthesis.

Pre-prohormone   A precursor protein that possesses an N-terminal signal peptide sequence in addition to the prohormone sequence.

Primer   A short oligonucleotide sequence that is paired with one strand of DNA and provides a start for the synthesis of a DNA chain.

Prion   An infectious agent thought to be composed solely of protein. The term is derived from 'Protein infectious agent'.

Prohormone   Inactive protein precursors which are activated by the removal of a peptide.

Prokaryotes   Lower organisms with no well-defined nucleus.

Promoter   Region of DNA molecule adjacent to a coding sequence where RNA polymerase binds and begins transcription.

Prophylaxis   Prevention of disease.

Protease (= Endopeptidase). Enzyme that catalyses the cleavage of peptide bonds in a polypeptide or protein. An exopeptidase is an enzyme that cleaves terminal amino acids from a polypeptide chain.

Protecting group Species which is used to mask a sensitive functional group while reactions are carried out on other parts of a molecule. When these reactions are completed the original functionality can be regenerated.

Proto-oncogenes A normal gene that, by alteration, can become an oncogene. The prefix 'c-' indicates a cellular gene, e.g. c-*myb*, c-*myc*, etc.

Recombinant DNA DNA molecule created by ligating segments of DNA that normally are not contiguous.

Repertoire The complete set of antigenic specificities generated by either B or T cells in response to foreign antigen.

Repressor A protein which regulates the transcription of genes under the control of an operator.

Restriction endonuclease An enzyme that recognizes and cuts specific base-pair sequences (restriction sites) within DNA.

Restriction site Base sequence recognized by a restriction endonuclease.

Retrovirus RNA virus which replicates via conversion into DNA duplex.

Reverse transcriptase An enzyme that catalyses the synthesis of DNA from an RNA template; this enzymatic activity is found in retroviruses.

Rheumatoid arthritis Autoimmune, inflammatory disease of the joints.

Ribonuclease An enzyme which degrades RNA.

Ribozyme An RNA molecule that possesses endoribonuclease activity.

Sense strand (of DNA) *See* Coding strand.

Serum Fluid portion of blood that remains after clotting has occurred; contains antibodies.

Serum sickness A hypersensitivity reaction, appearing 1 to 2 weeks after injection of a foreign serum, triggered by the deposition of circulating antibody–antigen complexes in tissue causing inflammation. The clinical symptoms are usually short-lived and resolution occurs spontaneously as the injected antigen is cleared.

Severe combined immune deficiency (SCID) Disorder resulting from reduced numbers of both B and T lymphocytes.

Signal sequence A 15–30 amino acid sequence at the N-terminus of a newly synthesized protein that serves to insert the protein into and transport it across a membrane.

Signal transduction Processes involved in transmitting the signal received on the outer surface of the cell (e.g. by antigen binding to its receptor) into the nucleus of the cell, which leads to gene expression.

Site-directed mutagenesis   Introduction in the test tube of specific mutation(s) into a DNA molecule at a pre-determined site.

Somatic cells   All body cells apart from the germ line.

Sticky ends   Short single-stranded and self-complementary sequences at the ends of largely double-stranded DNA.

Systemic   Relating to or affecting the whole body rather than individual parts or organs.

T cell   Lymphocyte concerned with cellular immunity.

T-cell receptor   A two-chain structure on T cells that binds antigen.

Template strand (of DNA)   Strand complementary in sequence of bases to mRNA transcribed from the DNA. Also known as the non-coding strand or the antisense strand.

β-Thalassemia   Genetically determined disorder caused by impaired synthesis of the β-polypeptide chain of globin, the protein part of the haemoglobin molecule.

Thrombolytic agent   Compound that breaks up blood clots.

Thrombus   Blood clot.

Tolerance   A state of immunological unresponsivnesss that prevents the immune system from mounting damaging responses against self-antigens.

Transcription   The process wherein DNA is copied into mRNA.

Transfection   The process by which viral or phage DNA is introduced into a cell or bacterium (a hybrid of transformation and infection).

Transformation   The process by which 'naked DNA' is introduced into a cell or bacterium. No other carrier substance is involved (cf. transfection). Also the cancerous alteration of mammalian cells.

Transgenic animals   Creatures that carry genes artificially inserted from other species.

Translation   The process whereby genetic information from mRNA is translated into protein.

Tumour-associated antigen   A protein that is found in tumours at significantly higher levels than in normal tissue.

Tumour suppressor gene   A gene whose product is involved in the regulation of cell growth. Prevents the transformation of a normal cell into a malignant one.

Upstream   DNA sequences before the start of a gene.

Vaccination   Protective immunization against a pathogen.

Vaccine Preparation of antigenic material that can be used to stimulate the development of antibodies and thus confers active immunity against a specific disease.

Vector A plasmid or phage into which foreign DNA may be inserted for cloning.

Virion A complete virus particle.

Virus A particle consisting of a core of nucleic acid (DNA or RNA) surrounded by a protein coat; it requires a host cell for replication.

Wild-type Genetic strain found in nature.

Index